CAMBRIDGE LIBRARY COLLECTION

Books of enduring scholarly value

Darwin

Two hundred years after his birth and 150 years after the publication of 'On the Origin of Species', Charles Darwin and his theories are still the focus of worldwide attention. This series offers not only works by Darwin, but also the writings of his mentors in Cambridge and elsewhere, and a survey of the impassioned scientific, philosophical and theological debates sparked by his 'dangerous idea'.

The Land of the Blue Poppy

In 1911, Francis Kingdon Ward (1885-1958) set off on his first solo expedition and collected hundreds of plant species, many previously unknown. From Burma, he headed into the Hengduan Mountains of northwestern Yunnan province, exploring along the Mekong, Yangtze and Salween rivers in the region between eastern Tibet and western Sichuan. In 2003, this area was designated a UNESCO World Heritage Site. One of the world's most biodiverse temperate zones, its extraordinary topography arises from its position at the collision point of tectonic plates. This fascinating book, first published in 1913, was one of the most popular by a prolific author. The blue poppy of the title is Meconopsis speciosa, which Kingdon Ward described as the 'Cambridge blue poppy'; rather than the famous 'Tibetan blue poppy' (Meconopsis betonicifolia) which he introduced to England in 1926. It is generously illustrated with Kingdon Ward's own photographs and maps from the trip, and will be read by armchair explorers and amateur botanists as well as by historians of science and those studying biodiversity in historical context.

Cambridge University Press has long been a pioneer in the reissuing of out-of-print titles from its own backlist, producing digital reprints of books that are still sought after by scholars and students but could not be reprinted economically using traditional technology. The Cambridge Library Collection extends this activity to a wider range of books which are still of importance to researchers and professionals, either for the source material they contain, or as landmarks in the history of their academic discipline.

Drawing from the world-renowned collections in the Cambridge University Library, and guided by the advice of experts in each subject area, Cambridge University Press is using state-of-the-art scanning machines in its own Printing House to capture the content of each book selected for inclusion. The files are processed to give a consistently clear, crisp image, and the books finished to the high quality standard for which the Press is recognised around the world. The latest print-on-demand technology ensures that the books will remain available indefinitely, and that orders for single or multiple copies can quickly be supplied.

The Cambridge Library Collection will bring back to life books of enduring scholarly value across a wide range of disciplines in the humanities and social sciences and in science and technology.

The Land of the Blue Poppy

Travels of a Naturalist in Eastern Tibet

F RANCIS K INGDON W ARD

CAMBRIDGE
UNIVERSITY PRESS

CAMBRIDGE UNIVERSITY PRESS

Cambridge New York Melbourne Madrid Cape Town Singapore São Paolo Delhi

Published in the United States of America by Cambridge University Press, New York

www.cambridge.org
Information on this title: www.cambridge.org/9781108004893

This edition first published 1913
This digitally printed version 2009

ISBN 978-1-108-00489-3

THE LAND OF THE BLUE POPPY

CAMBRIDGE UNIVERSITY PRESS
London: FETTER LANE, E.C.
C. F. CLAY, Manager

Edinburgh: 100, PRINCES STREET
Berlin: A. ASHER AND CO.
Leipzig: F. A. BROCKHAUS
New York: G. P. PUTNAM'S SONS
Bombay and Calcutta: MACMILLAN AND CO., Ltd.

Waterfall at T'eng-yueh, West Yunnan

THE LAND OF THE BLUE POPPY

TRAVELS OF A NATURALIST IN EASTERN TIBET

BY

F. KINGDON WARD, B.A., F.R.G.S.

Cambridge:
at the University Press
1913

𝕮𝖆𝖒𝖇𝖗𝖎𝖉𝖌𝖊:

PRINTED BY JOHN CLAY, M.A.

AT THE UNIVERSITY PRESS

TO THE MEMORY OF MY FATHER

HARRY MARSHALL WARD, Sc.D., F.R.S.
PROFESSOR OF BOTANY IN THE UNIVERSITY OF
CAMBRIDGE, 1895—1906

THIS BOOK IS AFFECTIONATELY DEDICATED

PREFACE

THE following chapters record my experiences and observations in Western China and South-Eastern Tibet during the year 1911, when I was engaged in collecting plants for the horticultural firm of Bees Ltd. Liverpool.

The sketch maps are drawn from Major Davies's map of Yunnan, with additions and corrections of my own. While the latter in no sense represent accurate surveys, I think that they will be of assistance to the reader, possibly even to future travellers. The photographs I took myself.

A great deal of the pleasure derived from looking back on a year's work in a distant land is associated with memories of the friends I made, and the help they so willingly gave me.

To Mr Archibald Rose, C.I.E., at that time British Acting-Consul in T'eng-yueh, I owe more than I can say. When I entered Yunnan in February 1911, the people were restless, and it seemed at first that I might not be able to proceed beyond T'eng-yueh. However, Mr Rose, having suggested A-tun-tsi as a likely centre for my plant hunting, promised to see that I got safely thus far, and did so. Not only did A-tun-tsi prove a first-rate collecting ground, but thenceforward everything went smoothly.

Mr E. B. Howell, Commissioner of Customs at T'eng-yueh, brightened many a lonely hour for me by forwarding newspapers and sending up my mails, which came, on the average, about once in five weeks; and to M. Perronne, a French gentleman who was buying musk in A-tun-tsi, and Mr Edgar, English missionary in Batang, I was also indebted for assistance.

Finally I must record the deep debt of gratitude I owe Dr Guillemard, who has edited the book and, during my absence, seen it safely through the Press. Dr Guillemard's great experience of travel not in Asia only, and his wide knowledge of natural history, were always at my disposal, an inexhaustible fund of fact and advice to draw on. Numerous alterations and valuable suggestions are due to him, and without the time he willingly expended on it, the work would have been much less presentable.

My one hope is that by the time these lines appear I shall be back again in 'the Land of the Blue Poppy.'

F. K. W.

CAMBRIDGE 1913.

CONTENTS

LIST OF ILLUSTRATIONS

MAPS

CHAPTER I

THE CALL OF THE RED GODS

On my return from Western China in September, 1910, I settled down to humdrum life with every prospect of becoming a quiet and respectable citizen of Shanghai. But in vain; travel had bitten too deeply into my soul, and I soon began to feel restless again, so that when after four months of civilised life something better turned up, I accepted with alacrity. This was none other than the chance of plant-collecting on the Tibetan border of Yunnan, and though I had extremely vague ideas about the country, and the method of procedure, I had mentally decided to undertake the mission before I had finished reading the letter in which the offer was set forth.

Three weeks later, on the last day of January, 1911, I bade farewell to my friends in Shanghai and started once more on my travels, sailing on the ill-fated *Delhi*, destined to make her last voyage just a year later. Soon we exchanged the bitter snow-storm which beat in our faces as we steamed out of the boundless Yang-tze for the warmth of the tropics, and I saw again the far-flung outposts of our eastern Empire, strung like gems at either end of that magic tiara of the Indies, which guard the approaches to the South China Sea. However fully the guardian islands of Hong-Kong and Singapore may satiate the inhabitants with their undoubted distempers, to the traveller at least they are never anything but charming.

At Penang, which in the business part of the city boasts nothing of beauty save an occasional Traveller's palm spreading its great fan over temple or hong—surely, as indeed the name suggests, one of the most remarkable of all the strange forms of tropic vegetation—I changed on to the British India boat for Rangoon, where we arrived three days later, and I spent a week in making preparations to go up country. It is not a really fascinating city, though the glory of the Shwe Dagon compensates for everything, and the gorgeous colouring of temple and lake, of earth and sky—here at last was the Oriental splendour of romance—rivets the attention of the newcomer. What it is all like during the south-west monsoon I do not know, but I imagine that the sunshine is all in all to Lower Burma. During the rains the dripping black skies must smudge the whole landscape with dreary greyness in spite of the vivid green vegetation springing to renewed life.

At last, my business completed, I entrained for Bhamo nominally three days' journey by rail and boat, and having a few hours to wait at Mandalay, I took the opportunity of visiting some of the sights in the ancient capital, a city of shops and temples. Thebaw's palace, now Fort Dufferin, to this day presents a crude but despoiled magnificence, mirrors and throneless daïses being the only conspicuous articles of furniture. After inspecting these and other glories of Mandalay in a sufficiently lethargic manner, for it was the hottest day I can remember, I returned to the station in time to catch my train.

Then northwards once more, past lovely meres where the wild-fowl wheeled in hundreds before settling down to rest, till the sun went down in a fog of crimson behind the purple hills, and we sped on into the darkness of another night. Early on the following morning we changed into the local train bound for Katha, on the Irrawaddy; and there we found awaiting us the steamboat which was to complete this tiresome journey to Bhamo.

There was little water in the river now, the spring rise not having commenced, but the fact that we ran aground in the middle of the afternoon and remained there till nine o'clock next morning, when we were pulled off, incommoded us not in the least; for it was all part of the journey and quite delightful after two nights in the train.

Late on the following afternoon we saw the white houses of Bhamo show up over the trees which fringed the river bank, and presently we tied up a couple of miles

below that curious little village of many vicissitudes. Nearer
we could not get, and the journey was completed overland
in a gharry. Having been in turn Chinese, Burmese, and
British, Bhamo could no doubt tell some strange stories of
frontier fights, raids, and other incidents of its chequered
career. Cooper, the great Chinese traveller, was murdered
here, and it was from Bhamo that the ill-fated Margary
started on his last journey. In addition to the British
authorities, civil and military, the polyglot population of
Bhamo now includes Burmans, Chinamen, Shans, Kachins,
Chittagonians and other peoples from India, while specimens
of most of the frontier tribes are occasionally to be seen
there, and a large volume of trade still passes through the
little border town in spite of the French railway to Yunnan-
fu. At this time it was probably more lively than usual,
on account of the friction on the Burma-Yunnan frontier
further north.

I spent five days in Bhamo, chiefly waiting for some
of my luggage which the railway company had failed to
account for satisfactorily, and the only civil authority with
whom I had dealings was by no means encouraging and
from start to finish poured cold water on my proposed
journey. My preparations, however, being as far as possible
complete, I decided to delay no longer in Bhamo, but to
cross the frontier at once. I had obtained the services of
a civilised Kachin to minister to my needs until I could engage Chinese servants to go with me the whole journey, but
at the last minute his wife put her foot down (even Kachin
women can assert themselves in a crisis) and refused to
let him come. I was therefore abandoned to the tender
mercies of a somewhat unlovely looking lad of doubtful but
decidedly mixed parentage and little experience, who nevertheless served me faithfully as far as T'eng-yueh.

What delightful fellows the Kachins are! I was quite
distressed at parting with my tribesman. Clean, neatly
dressed, and debonair, he stood waiting motionless behind
my chair or moved noiselessly round the bungalow as a
waiter moves round a first-class London club.

On February 26 the mules were loaded and headed
towards the distant dancing hills; an hour later I too
mounted, and turning my back on sun-scorched Bhamo,

cantered slowly down the long white road that leads to China.

The initial stage out of Bhamo is only nine miles, and it was undoubtedly this fact alone which caused me to feel extraordinarily lonely on the first evening of my journey. Arriving very early in the afternoon there was of course nothing to do but to take out a gun and look round for game, but, do what I would, there was no getting away from the sense of utter desolation which seemed to crush me. Even the mild excitement of putting up a barking deer amongst the reeds of the river failed to alleviate the depression and after dinner I was only too glad to crawl into bed and, weary in spirit, court oblivion in sleep. Never again did the sense of paralyzing isolation come so vividly upon me as on that first night, when all the trials that awaited me seemed to take shape and rise in arms to mock my ignorance and feebleness.

The scenery as the plains of the Irrawaddy valley are left behind and the road gradually ascends the mountain side to traverse the gorge of the Taping river, grows more and more picturesque, and the booming of the torrent, soon a thousand feet below us, alternately dies away and swells up louder and louder as the road sweeps round the gullies. Finally we catch a glimpse of it foaming over the rocks, and then it quickly dies out of sight and sound once more, till only the tinkle of our gongs echoes through the slumbering forest. The mules, with the natural cussedness of the breed, trudge stubbornly along on the extreme edge of the precipice, though the road is, as a matter of fact, respectably broad here. It takes a little time to get accustomed to the idea of riding along with one leg hanging over the edge of a precipice, whence a sheer drop would land one on the tree-tops hundreds of feet below.

On the fourth day we crossed the bridge which marks the frontier between two Empires. To us in our little island, a frontier sounds a more or less nebulous quantity, something drawn rather whimsically on maps, and a chronic source of petty international jealousies as difficult to define as the boundary line which gives rise to them. But this elusive idea becomes almost a physical reality when one crosses the frontier of a British possession overseas,

Map I

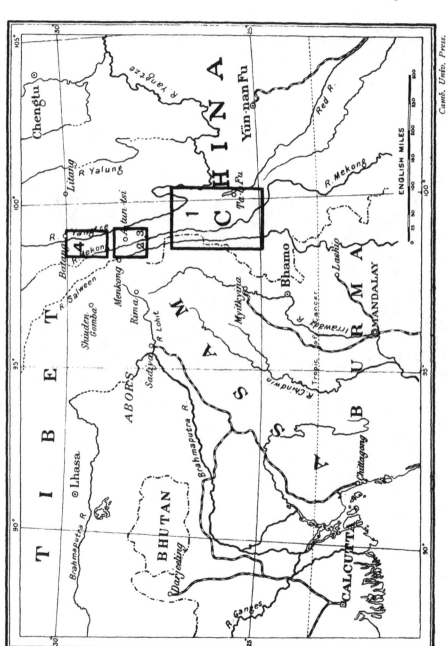

Key Map. Showing the relation of the region traversed to China, Tibet, Burma, and Assam.

[*The figures refer to the large scale sketch maps.*]

Camb. Univ. Press.

thus bringing to a focus, as it were, the days which are past and all that lies before one in the new world. Especially is this the case on the return journey, when the hardships are over. Never shall I forget the thrill of joy which quickened me when I crossed the Yunnan-Burma frontier on January 1, nearly a year later, and looked back down the vista of months spent far from our heritage in the East. It was not that the future seemed much brighter than the past, for never had I enjoyed myself more; not that I found the efforts of a Public Works Department— erect telegraph poles and taut wires, reliable bridges, mile posts, and rest-houses provided by a paternal government— filling a long-felt want; but simply that the act of crossing our own frontier again, with all that that frontier stood for, made my heart throb a little more quickly.

A few miles farther I was surprised to see an Englishman sitting in the doorway of a hut on the mountain side, smoking a pipe, and closely watching some fifty coolies who were busily engaged in mending the road.

I of course stopped for a chat and soon discovered that my companion was a keen naturalist, years of lonely watching while engaged on such work as this having made him extraordinarily observant, and quick to detect the slightest movement. Mr Oliver, for such was his name, asked me if I had seen any monkeys, and on my replying in the negative, he merely said : " Then watch that tree."

I looked down the mountain slope in the direction indicated to a strapping forest giant that spread 'aloft a great canopy of branches hanging above the road, and waited. One minute, two minutes ; not a leaf stirred ; the forest was silent and seemingly deserted ; not even the tinkle of a stream disturbed the profound quiet. And then suddenly, as though a breath of air had sighed over the jungle, a shiver seemed to pass through the branches of the tree, and almost immediately a brown shadow appeared out of the foliage, ran along a branch which swayed dizzily, and crouched ; he was followed by another, and another, and yet others, now plainly visible, and still the branch swayed rhythmically as it became more and more depressed. "Phayre's leaf-eating monkey" (*Semnopithecus Phayrei*), said my companion shortly; "watch them travel from tree to tree."

At that moment the first monkey leapt; there was a splash of foliage in the tree below, the branch, lightened of a portion of its load, recoiled, and as it came down again the second monkey sprang into the air, hands and feet neatly gathered together. And so they went on, the monkeys leaping one by one from the branch end as it swung up and down, till there was but one left, the branch by this time see-sawing to within a short distance of the goal. Down went the last of them, tail streaming out behind, and away into the jungle after his companions, who had by this time swung themselves out of sight.

But it was now time for me to move on in pursuit of my caravan, and leaving the naturalist to superintend his coolie gangs, I went my way, eyes very wide open, trying to see more monkeys.

No event worth recording took place till the fifth night, when the harmony of the evening was temporarily interrupted owing to the exceedingly bellicose attitude of the innkeepers at Chiu-cheng who, with unusual singleness of purpose, one and all refused to admit me. Eventually I was driven to seek shelter beneath the roof of the village school house, and can testify that here they look upon education with an indulgent eye, since the fact that I might annex endless school books did not weigh heavily with them in comparison with the fact that they did not desire my presence within their homes. It occurred to me that I was now beginning to feel the full force of Chinese displeasure over the Pien-ma incident on' the frontier, but subsequent events caused me to modify this view considerably.

By this time we had left the river gorge and the teak forests behind us, and down in the open valley, where the Taping flows between extraordinarily bare treeless hills, we had rain, and the road rapidly resolved itself into a quagmire, which the peasants were diligently adding to by dredging their rice fields and dressing the track with the semi-liquid slime.

It is but eight mule stages from Bhamo to T'eng-yueh, and the road is sufficiently well known to require no detailed description here; but I was destined to meet with a small mishap on the eighth day when, starting early in the morning and leaving the mules to follow at leisure, I dashed off alone and lost the way. However, after following devious paths somewhere south of the main pack-road, I ultimately reached the city late in the afternoon, though not before my mule had thrown me three times, and I was thoroughly hot and exasperated. Curiously enough, though starting on the right road at the other end, I lost my way again on this very same stage from T'eng-yueh ten months later, finding yet a third route with considerable success—of a sort; so that I have still to discover the proper road over this section.

Arrived at the city, I marched straight into the Consulate and surprised almost the entire European population of six having tea with Consul Rose, who, in spite of my dishevelled appearance, gave me a very warm welcome.

While it is undoubtedly true that I had come into Yunnan during a period of stress, the continued forays over the frontier into the Kachin country of Upper Burma having led to a British expedition in that direction, things were not so hopeless as the Deputy Commissioner in Bhamo had painted them. But in any case I now had the Consul at my back, and a short chat with him was enough to dispel any suspicion of gloom which might have tended to come over me when I reviewed the prospects of success. Mr Rose suggested A-tun-tsi as likely to prove an excellent centre for my work, promising that if I found any difficulty in getting there, he would take the necessary steps on my behalf. The Taotai, indeed, was an altogether wretched person, anti-foreign by nature and furious with the British on account of the frontier trouble; but his position between the devil and the deep sea was by no means an enviable one, for Mr Rose had already brought pressure to bear on him owing to a local boycott of British goods. The consequence of this, as he informed the Consul with a wry smile, was that people from all over the province had written to him, cursing him for showing favour to the British, and no doubt he would have liked to stop me from going further into Yunnan. Realising that the futility of such an action would have been made abundantly clear to him, however, he took a safe line and gave me permission to go to Lichiang-fu, questioning the Consul

closely as to what I was doing.   No mention was made of
A-tun-tsi, for it is a safe thing in China not to ask too
much of any man, but to go from one to the other, approaching the more friendly officials and ignoring others
according to circumstances.   To the Taotai at T'eng-yueh,
A-tun-tsi was probably a savage place where people were
engaged in cutting each others' throats and a European
would infallibly be killed, whereas to the official at Wei-hsi
it was the obvious place to make for.   Once in Lichiang
I was quite beyond the control of the Taotai, though not
out of reach of the Consul, so that this concession was
perfectly satisfactory as far as it went, though I had no
intention of going to Lichiang.

I spent twelve days in T'eng-yueh, waiting for my
baggage, as the guest first of Mr Rose, then of Mr Howell,
Commissioner of Customs, and delightful days they were—
scampers over the grave-strewn downs on the spirited little
Yunnan ponies, snipe shooting, and occasionally a game of
rounders with the 'boys,' cooks, gardeners, and other
members of the several households.   Those games of
rounders in a little dell surrounded by the necropolitan
hills were great fun, for the Chinese were as keen as
schoolboys on the game, many of them after a little practice
showing surprising skill.   But it was a little disconcerting
in the middle of an exciting international match when
seven o'clock came and the ranks of both sides were
suddenly decimated by the defection of the cooks, and the
cooks' boys, and the cooks' boys' helps, who all rushed
frantically away to prepare dinner.

At this time there was a little flutter of excitement
amongst the half-dozen Europeans at T'eng-yueh, owing
to the forthcoming marriage of the 'General's' daughter.
The 'General' was a man who had been sent to pacify the
T'eng-yueh district during the great Mohammedan rebellion
fifty years before, and had found it so pleasant that he had
stayed there ever since.   He was a great favourite with the
Englishmen, who were amongst those who sent wedding
presents, and it is sad to recall that he was one of the first
victims of the rebellion seven months later.   According to
current gossip the sponsor for his daughter's worldly goods
was only accepting one thing in four, a course which while

Plate I

Temple built over a boulder
near Tali-fu

Mr A. Rose, C.I.E., Late Acting-
Consul at T'eng-yueh

allowing plenty of latitude to his English friends now busy racking their brains to find acceptable gifts, also implied that the old man enjoyed exercising his prerogative of choice. The Consul came out strongly with a bottle of scent, a box of Vinolia soap, and a silk handkerchief. The fourth item was rather a puzzle, but not to be daunted a bright idea finally came to him, and having filched a nice blanket with pretty blue and red stripes from his bed, he despatched it on a tray with the other articles; then, remorse succeeding, he spent the afternoon in an agony of apprehension lest the old man should choose the blanket, and he be compelled to freeze every night till he could get another one up from Bhamo. But this was as nothing compared with the horror which presently possessed him. With a view to enhanced effect the Consul had covered over the things on the tray with his best tea-cloth, and no sooner were the wedding presents under way than an appalling thought occurred to him. Suppose the old man, instead of whisking off the table-cloth to gaze with rapt eyes on the treasures beneath, as was expected of him—suppose that his gaze should be arrested before the curtain went up, so to speak, and that with nice discrimination he should choose the tea-cloth itself before he had a chance of falling in love with the scent or the soap! The Consul groaned at the bare possibility; it was just the sort of tactless thing a retired General might be expected to do.

However, the old man rose grandly to the occasion, eschewed the table-cloth and the blanket, and chose the silk handkerchief. The Consul slept warmly that night, we spread the tea-cloth again next day, and all came back from the wedding feast, which was a sort of high tea lasting from 5 p.m. till 9 p.m., looking pale but satisfied.

On March 15th my missing baggage came up from Bhamo and on the 18th I started eastwards, having in the meantime procured the services of a Minchia cook, Hoshing by name, who had had some previous experience with Europeans.

Sorry as I was to say good-bye to my friends in T'eng-yueh, who had done everything to give me a real good time, I nevertheless felt in the best of spirits at the prospect of leaving. The sky was turquoise blue, and the

snow which flecked the high peaks of the black ridge
between the Shweli and Salween rivers glittered in the
sunshine.   The rolling hills of the plateau still looked bare
and brown in spite of early spring rains, but riding slowly
over them, one found brilliant gentians peeping up from
the grass, purple orchids and white dog-roses in the lanes,
hedges of scarlet *Cydonia japonica* broken here and there by
masses of pink peach-blossom, and everywhere green leaves
unfurling.   In short, a joyous note of spring was already in
copse and spinney when I finally set out on my journey.

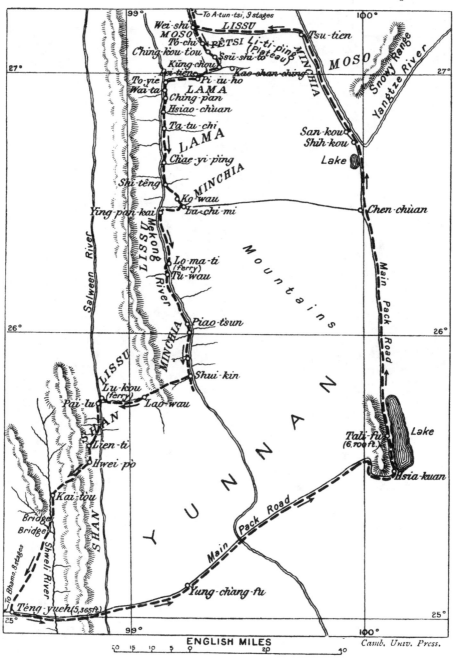

Map II

Sketch map to illustrate the route from T'eng-yueh to Wei-hsi, and the return journey through the tribal country. [*Chapters II and III; and XV and XVI.*]

CHAPTER II

ON THE PLATEAU OF YUNNAN

THE same incidents with only a background of varying details fall to the lot of most travellers in the interior of China, and I recall nothing very remarkable during the fourteen days which elapsed between leaving T'eng-yueh and arriving at Tali-fu. The Ambassadors' Road, as it is called, has been well described many times, though it is interesting to compare one description with another and to note how very different the same journey may appear to people whose interests in life are more or less different.

As for me, I have an eye for plants, and take more than a passing interest in men and things; and to my mind the high plateaux between the deeply-scoured, trench-like valleys, now blazing with scarlet rhododendrons and pink camellias, afforded such charming landscapes that I was almost oblivious of everything else.

After the mid-day halt, I would leave the men to load the animals, and taking my gun, turn aside and wander alone amongst the park-like undulating hills, finding here sheets of mauve primulas blooming on emerald grassy slopes, dog-roses yellow and white, pale-blue irises, and other delightful flowers. Flocks of green parrots flashed screeching overhead, seeking the red berries of a species of mistletoe that grew on the pine trees; gorgeous little fly-catchers flitted timidly from bush to bush; and sometimes I would put up an Amherst's pheasant, perhaps the most magnificent of the tribe, with its handsome tail and rainbow neck. Down by the stream one might generally find a David's squirrel frisking amongst the rocks, but I saw no other mammals, nor would one expect to by day.

At the insignificant village of P'u-p'iao quite a disagreeable incident interrupted the monotony of the journey, for the villagers exhibited the curiosity of impudence to such a degree while I was putting up my tent, that I had a row with one of them.

Meanwhile quite a crowd had collected to see the fracas, and it says much for the good sense and peacefulness of the Yunnanese that they did not take sides. Possibly a few, a very few, thought the man deserved what he got ; still fewer might have been really afraid of a foreigner ; but undoubtedly the vast majority were supremely indifferent— it was none of *their* business.

In the night my tent was robbed. At the head of my bed stood a small table on which was a thermos flask full of hot tea, to be taken first thing in the morning while dressing, and a towel ; I had left the front of the tent open for it was a warm night, and as I lay in bed I was thus able to watch the moon rise over the ebony range of mountains to the east. Next morning the table was bare ; my thermos flask and towel were gone. Whoever the thief was, he had simply put his arm through the tent opening and taken what he could reach without any trouble or noise. Suspicion pointed strongly to my friends of the previous afternoon, for it was unlikely that a casual tramp had come along and walked off with the things—his approach would be heralded by the barking of the village dogs ; besides, a beggar could not go about in China with a towel over one arm and a thermos flask under the other without attracting considerable attention.

It was all very annoying. The towel did not matter so much, for I had others, but the thermos flask being one of those small luxuries which add enormously to one's comfort on such a journey was a serious loss ; nor was that all, for it had been a parting present to me from my boys at the Shanghai Public School, and I much resented losing it. However, nothing could be done till we reached Yung-ch'ang that same evening, when having made myself respectable, I rode round to see the official, preceded by Ho-shing bearing my card and passport. Having ridden through the first two courtyards of the Yamen I dismounted before the closed inner gates and waited, while retainers seemed to spring up on every hand from the very paving stones. A few minutes later somebody called out in a loud voice that the great man was ready, the gates were flung open, and I marched across the inner court and up the steps to where the mandarin, a kindly-looking grave-faced man with delicate hands and long moustaches, stood waiting to receive me; I bowed low to him, and acknowledging the salutation he ushered me into a small room, plainly furnished in excellent taste if not scrupulously clean. In the centre of the apartment was a round table covered with heavy red cloth, and at the far end a low daïs with two cushioned seats upholstered in similar fashion. Set against the side walls in severe symmetry were two small square tables, in some dark varnished wood, flanked by chairs to match, all rather gawky, stiff, and uncomfortable; and the walls themselves were hung with long scrolls, some bearing crude sketches of scenery, others displaying proverbs or quotations from the classics. The trellised window frames were covered with tough translucent paper, letting in plenty of light for ordinary purposes, besides plenty of air. That was all—only the floor was of rough boards, neither varnished nor carpeted, very dusty and bearing signs in the shape of burnt matches, nut-shells, cigarette ends and numerous expectorations, of previous visitors. It was a typical guest room of a native Yamen. "Please be seated" said the Prefect motioning me to the left-hand seat on the daïs—the seat of honour, and himself taking the right hand, while Ho-shing sank into one of the seats of dishonour and discomfort against the wall lower down the room, and the retainers stood in an expectant knot round the door. Ho-shing then explained the object of my visit, and after the Prefect had asked a few questions, I lifted the cup of tea which had been brought me in token of departure, rose, and took leave.

The only practical result of the interview was that I had to spend a day in Yung-ch'ang while two soldiers returned to P'u-p'iao to recover my lost property; and needless to say they returned the same evening empty-handed.

It is the misfortune of China that the innocent frequently suffer with the guilty, since it is the crime, and not necessarily the criminal that is punished. While therefore it is probable that retribution of a sort did fall, as I had prophesied, on the wretched village of P'u-p'iao, it is certain that I did not gain by the transaction, and we may surmise that the soldiers, unwilling to take back a negative report to the mandarin unless the villagers made it well worth their while, reaped the only benefit.

Next day I found myself furnished willy-nilly with three 'braves' as escort, though they were armed with nothing more formidable than a fan and a water-pipe between them. These fellows as a rule strolled comfortably along in the rear, yelling officiously whenever we met other mules, and thoroughly dislocating the traffic, though their avowed object was to clear the road for me.

The plateau of Yunnan is scarred from north to south by deep trench-like valleys, at the bottom of which flow the Shweli, Salween, Mekong, and other less known rivers, all crossed by means of chain suspension bridges which have been described by previous travellers on this road. The astonishing difference between the comparatively broad forested Salween valley, and the narrow rift, its stark cliffs almost completely devoid of vegetation, through which the Mekong flows, is only a foretaste of what is to come later in the north.

After the Mekong came the Shun-pi river, and drenching rain for two days. As we climbed up into the mountains again, here of red and green porphyry, it was interesting to note how the gullies where the torrents had their source, instead of contracting as the streams grew smaller, opened out more and more, their funnel-like mouths choked with débris, while great fans of rough-hewn rock had been flung out athwart and around the now puny stream. This was good evidence of the furious summer rains which descend upon the mountain summits, sweeping everything before them.

The alders and birches were bursting into leaf now, and there were new blossoms on the road—barberry bushes with tall pyramids of yellow flowers (*Berberis nepalensis*), and deliciously scented white jasmine. Sometimes we met children carrying balls of these delightful jasmine flowers cut off short and tied up in this way, so that each ball dangled from a thread, a solid sphere of scented flowers, and the muleteers bought them to twist round their buttons, or decorate their hair.

At the village of Yang-pi we crossed a fair-sized river by the fifth and last chain bridge on this road. There must be quite a large number of these remarkable structures spanning the smaller rivers of Western Yunnan—I crossed two over the Shweli alone, but of their history I know nothing, though they have doubtless been in use from very ancient times.

Continuing down the Yang-pi river next day, we saw at last the snow-capped peaks of the high range over-hanging Tali-fu. After the capitulation of Tali at the close of the Mohammedan rebellion sixty years ago, many fugitives from the massacre tried to make their way from the rivers of blood which flowed in the doomed city across the white snows of these jagged spires, but the majority perished miserably of cold and hunger before a practicable route down to Yang-pi was discovered.

In the afternoon, while scrambling about by the Yang-pi river, I came across a remarkable bridge of twisted lianas, like those built by the Kachins and other jungle tribes. It was really a suspension bridge, the two main liana cables being fastened to stout bamboo poles on either bank and supporting a shallow hammock made of similar lianas inter-laced in skeleton fashion, so flimsy that a single narrow path of six-inch planks had been laid along the middle line to mark the fairway. I cannot imagine a more unstable bridge worthy of the name. Not only did it sag at least six feet in the middle—it was barely thirty-five yards across—but the least breath of air set the entire structure swaying in the most sickening fashion, and what was worse, no sooner did one set foot on the planking than the hammock began to wriggle beneath one's tread, partly this way, partly that, so flexible was it. The only way to negotiate this bridge in safety, and not tumble over the side or fall through the basket-work, was to step gingerly along the line of planks sliding one foot carefully in front of the other all the time, while striving to maintain a good balance, in spite of the river raging over the rocks thirty feet below, the roar of which filled the air.

Presently three women, each bearing a heavy load,

came down the river bank and crossed one at a time whereupon I was filled with ambition to walk on the crazy bridge myself, and promptly essayed the journey, though before I was half-way across, I repented of my rashness. I did not dare turn round, however, so I had to go through with it, though I felt more like going down on all fours and crawling than standing upright. I wished there was a handrail to hold on to, and that the bridge would only stay still a minute! The worst of it was that I had to return!

On March 31, we reached Tali-fu, and there being no mules immediately available, we stayed there a week. It was a little tedious, but it afforded an opportunity of seeing the beautifully situated city under varying moods of sunshine and storm—black billows of cloud surging over the snow-tipped peaks to the west, the great blue lake lying like a sheet of glass in the sunshine, or sullen purple as the sky darkened and the wind sent foam-crested waves rocking up the beach; the mountains across the water, golden-yellow and red in the morning, crimsoning as the sun went down.

A few miles from the city is a fine temple built over a boulder probably rolled down from the mountains above, though local legend accounts for its origin on the narrow plain in a much less prosaic manner. I do not know the complete story, but Kwan-nin, the Goddess of Mercy, was apparently responsible for bringing it along and dumping it down here, wherefore it is not surprising that, in the old religious days, a temple should have been built on the spot to commemorate such a miraculous deed. A picturesque shrine, on which is inscribed the whole story, forms a canopy over the boulder within the temple court, and several cypress trees add a touch of sadness to the old neglected building.

One day I walked on the city wall, where I met a smartly dressed Chinaman carrying a pet bird in a cage which he had brought out to take the air, as the Chinese are wont to do; and to my great surprise this Celestial addressed me in French, pronounced with a perfect Parisian accent. Recovering from my momentary amazement I was about to reply when I suddenly discovered that I did not

Plate II

Street and Gateway, Tali-fu

know one word of French, or at least I could not put my tongue to it! It is very curious how one may grope for the simplest words in a language never thoroughly mastered in the past when one is learning a second language and hearing it spoken daily. It was annoying to think that I could have answered his questions in Chinese, but not in French; while on the other hand had I answered in Chinese I should have been still further humiliated when he spoke in his own tongue. Certainly the greatest difficulty in acquiring a language is to have the words instantaneously ready for use, and this can be accomplished only after long practice amongst the natives, never from books.

At last I resurrected a few half-dead words and attempted a remark, but it was a deplorable failure and he corrected me with a faultless accent. Never had I felt more ignorant of French! And then "Vous êtes Anglais! Je ne parle pas Anglais"! he said apologetically, shaking his head, but I did not believe him and kept quiet. He might have corrected my English too had I spoken in a hurry.

This man was a Catholic and had been taught French by the priest in Tali. I sometimes wish our own missionaries would take the trouble to teach their evangelists English.

At Tali I obtained the services of two more men, Kin and Sung, Chinese both of them, who were destined to follow my fortunes for the rest of the year, Kin in particular doing excellent service. On April 7 the mules were to hand, and we started off again, journeying northwards.

CHAPTER III

ON THE LI-TI-P'ING

NORTH of Tali are bare rounded hills of red earth, and richly cultivated plains tucked in amongst the mountains where, mile on mile, wave fields of kidney-bean, wheat, and blue-flowered flax.

Sometimes we would meet strings of women carrying loads of salt, cotton, beans, or rice to the local market and driving mules laden with planks and firewood. Curiously enough these women supported the loads on their backs by means of a strap passing round the forehead, after the manner of jungle tribes and dwarf races, thus walking with bent backs and contracted chest; and certainly, except for their clothes, the people about here had nothing Chinese in their appearance, being mostly Minchia with a very pronounced type of countenance.

One evening we came upon an isolated limestone hill, curiously sculptured into holes and caves, and from its base issued two very hot springs smelling strongly of sulphurous gases. Such springs are abundant in Western China at the foot of every great mountain range. On the fourth day after leaving Tali we reached Chien-ch'uan, an important market city standing at the head of a small plain, partly occupied by a lake, from which rises the Yang-pi river. Continuing northwards, the undulating valley began to take on more and more the character of plateau country, the ascent being very gradual, albeit we were hemmed in by mountains on either side, those to the east still capped by winter snow. There was little cultivation now, the valley floor being frequently boggy and used chiefly for grazing purposes, while the uncleared mountain slopes were covered with brushwood below and with pine forests higher up.

Abruptly came the head of the pass, and the narrow plateau seemed quite suddenly to drop away into a big blue valley, in the depths of which, down beyond the forests, lay hidden the Yang-tze, or as it is called locally, the Kin-sha, the famous River of Golden Sand. Far beyond there towered into the now cloudless sky the Lichiang range—a magnificent snow-clad group of mountains which have barred the way of the river and thrust it aside, thus causing it to sweep round in a vast bend to the north, and return almost on itself in a narrow loop which adds hundreds of miles to the length of the Yang-tze, before it finally sets out on its long journey to the China coast. Up and up towered those glittering pinnacles of ice and snow, flashing in the sunlight till they melted away into the infinite blue. It was a glorious sight.

Here at the very summit of the pass, over 8000 feet above sea-level, a beautiful little lake nestled close under the high wooded cliffs of the eastern ridge, owing its origin to an alluvial cone which had swept across the valley lower down and dammed back the water.

This watershed separates the streams flowing directly to the Kin-sha from those flowing to the Yang-pi river, itself a tributary of the Mekong. Not far to the north, the three great rivers, the Salween, Mekong, and Yang-tze, flow in their parallel gutters. The descent to the deep-lying Kin-sha was extremely precipitous, the valleys on this side having been torn out of the mountains by furious and sudden rains. However, it was not far to the village of Shih-kow, situated by the river, and here we halted for lunch, continuing in the afternoon to San-kow.

The Yang-tze or Kin-sha at this point, 2500 miles from its mouth, averages perhaps a hundred yards from bank to bank, and though at this season of low water the stream is much subdivided by islands of sand and shingle, it runs with a strong current. However, it was already the second week in April, and the snows had begun to melt in the far north, so that in spite of the sunshine the crystal water was bitterly cold, as I had good reason to know; for having shot a Brahminy duck which had settled down on an island, I was compelled to strip off my clothes and swim across a narrow backwater in order to secure my meal. When summer comes with its continuous drenching rain on the Tibetan plateau, a vast brown flood of water rages down here, filling the broad channel from bank to bank, submerging the islands, and sweeping everything before it. Then, as the fine autumn weather sets in and the iron frosts lock up the mountain torrents again, the water falls gradually, growing clearer and bluer, the duck return to the islands, and the gold washers come back to their re-sorted gravel pits. It is only another phase of the far-reaching monsoon.

The closely investing mountains rise several thousands of feet above the river, but the valley is sufficiently broad to allow of considerable cultivation on the right bank, where a platform averaging a quarter of a mile in breadth separates the river bank from the mountain foot. No such facilities exist on the left bank, however, and villages on that side are practically confined to the wide-mouthed breaches opened out by tributary torrents. At this season there are no rapids to speak of, but the swift current makes the river useless for navigation except locally, where ferries ply across, and fishing is carried on from large scows.

As to the people who inhabit this stretch of the Kin-sha, they seem to be mainly a cross between Chinese and Tibetan, with a considerable admixture of Moso blood from the immediate north and Minchia blood from the Mekong valley to the west. The women have peculiarly broad faces, which give them a merry, good-humoured expression. They are fond of wearing small ear-rings consisting of a silver ring, like a broad finger ring, from which is suspended a jade disc, pierced in the centre; otherwise they affect little ornamentation.

From day to day we passed groups of people washing for gold, the method of procedure being as follows:

The gravel at some chosen spot exposed during the dry season is shovelled into a basket, which is rocked by hand on the edge of a long inclined sieve, water being poured in to wash the fine mud through the interstices of the basket, the remaining stones being flung aside. As the mud and water trickle down the sieve, the water drips

Plate III

Washing for gold in the bed of the Kin-sha

Lake scene, Yunnan plateau

through and the yellow mud collects below, whence it is taken to the river side to be rocked in a shallow wooden cradle. The light mud is soon suspended in the water and gradually slopped over the side, leaving at the bottom of the cradle a heavy black sand, in which lie the still heavier flakes of gold. This black sand is cradled more carefully, and finally we have the specks of gold alone glittering in the cradle.

Operations are best carried on by seven or eight persons, one half being engaged in extracting the auriferous sand, which lies usually a few feet below the surface shingle, the other half cradling the sand; but sometimes there are only three or four people to a syndicate, men, women and children being employed indiscriminately, though of course it is not particularly arduous work. The earnings vary considerably, but a hundred and fifty cash (about $7\frac{1}{2}d$.) a day per person seems a fair average.

I have never seen this business carried on north of Pang-tsi-la, though it must be remembered that when I next saw the Kin-sha the water was at summer level. However, near Batang facilities for gold-washing certainly existed, and I believe I am right in saying that it is never undertaken; consequently I am inclined to ascribe a local origin to the gold.

On the Salween gold-washing seems to be entirely unknown, though I heard of it near the headwaters of the Irrawaddy, and on the Mekong I only once saw a few abandoned diggings; hence there is little ground for supposing that the gold originates from big reefs far up in Tibet, otherwise one would expect to find auriferous sand in the Mekong and Salween. Still I am inclined to think that, in this localised area, gold exists in considerable quantities along the Kin-sha, for even with these primitive appliances the men washed sufficient gold from the mud for me to see in a few minutes.

Nuggets are rarely found, and for obvious reasons, since they must be for the most part at the bottom of the river. Dredging or some more rapid and thorough treatment of the auriferous shingle would doubtless be a profitable undertaking, but it is not likely that innovations of this sort would be viewed with favour by the local families who

have, or at all events imagine that they have, vested rights
in the river bed.

On April 14 I decided to give the men a day's rest. We
had pitched our camp in a delightful grove of trees, just
above the river, and the day being fine, Kin and I set out
to climb to the summit of the ridge immediately west of us.
However, after reaching an altitude of over 10,000 feet
progress was stopped by the limestone precipices which
crowned the ridge, and finding deep snow still lying in the
corries, and no sign of spring as yet, we returned to the
warmth of the valley.

On the downward journey we came across some
pheasant springes of a kind seen in many parts of the
world—a running noose attached to a bent bamboo held
down by a catch, which on being released at once flies up
and suspends the struggling bird in the air.

Next day, our last by the Kin-sha, we went on to
Chi-tien, a semi-Tibetan village of no importance, passing
through groves of Chionanthus trees in full bloom, a
glorious sight. They seem fond of the rockiest places
down by the river side, and are often to be seen hanging
far out over the water.

For the last three days we had been passing a good
deal of opium poppy—not solid fields of it such as I had
seen in one or two places between T'eng-yueh and Tali,
even on the main road, but scattered plants occurring
amongst fields of white and mauve peas, which served
admirably to mask their presence, and there can be little
doubt that these poppies were deliberately sown amongst
the peas so that seed might be kept against better times.
Under an able and. powerful Viceroy Yunnan, like several
other provinces of the Empire, had almost rid itself of
opium, but before the work was completed, there had been
a change in the administration, the present Viceroy being
not only a weak ruler, but an opium smoker himself ; hence
poppy cultivation, though still showing a substantial de-
crease since the opium edict, had made a considerable
advance again during the preceding season. It is hardly
fair, however, to compare Yunnan with any other opium-
growing province, since the conditions are totally dissimilar.
The Yunnan plateau is by no means suitable for all crops,

cotton in particular doing badly, and there can be little doubt that the poppy is more suited to the climate than almost any other economic plant. Consequently the greater part of the opium had in the past been sold to Cantonese and Hunnanese merchants, cotton cloth from Ssu-chuan and Kwei-chow being bought with the silver so obtained. The reckless extermination of the poppy had therefore involved the province in serious financial difficulties, the shortage of silver and the fallow fields—for it was then too late to plant anything else—creating much misery.

Again, in the deep valleys of the plateau malaria is rife during the rainy season, and the Shans and other tribes who inhabit these valleys use opium as a prophylactic; for it must not be forgotten that, except along the main roads, there are very few Chinese in Western Yunnan, and though many of the tribes have been more or less absorbed into the dominant race, such fundamental customs as the eating of opium to insure immunity against fever are not easily shaken. It was so used in the Cambridgeshire fens until quite recently. Western Yunnan probably received opium from India long before the poppy was grown in China.

Finally, opium being extremely light, and at the same time acceptable to almost everybody, it forms a convenient medium for exchange in a sparsely populated country of immense distances and few roads, and is commonly used as such in place of silver. Bearing all these things in mind then, it will readily be admitted that the Yunnanese had a grievance when the extermination of a plant which meant clothes and medicine to them was attempted. How the present revolution will affect poppy cultivation in the more populous and accessible provinces it is difficult to say, but it is a foregone conclusion that with the setting up of a powerful local authority, there will be another large increase in the cultivation of opium in Yunnan this year.

The journey westwards from the Kin-sha over the Mekong watershed to Wei-hsi-t'ing, whither we were now bound, takes two days under ordinary circumstances. Leaving Chi-tien, we ascended a small valley, cultivated below, but presently forested, the stream being in many places jammed with logs which had been cut in the forests above and slid down the steep slopes to make the best of their way to Chi-tien. On the second day a stiff climb in a heavy snow-storm brought us out of the forest on to the Li-ti-p'ing, as the summit of the watershed is called—a desolate plateau of grass-land, forest, and bog. Streams of discoloured water trickled sluggishly down the grassy hollows, but the hill-tops were covered with fir forests, where patches of snow still lay snugly hidden. A raw wind swept over the pass and down the shallow valleys.

It looked an ideal spot for pheasants, and leaving the caravan to continue across open country, Kin and I climbed the slopes to the edge of a forest patch. Shortly afterwards Kin complained of feeling unwell, so taking my gun I told him to rejoin the caravan while I scrambled about by myself, and presently they were all hidden from view. For about an hour I wandered from one patch of forest to another marvelling at the reckless slaughter going on in the plant world around me. The damp and darkness of the fir forest with its bamboo undergrowth was a wonderful breeding-ground for moss, while coils and coils of pale green unhealthy-looking lichen flung insidious tentacles round every tree, slowly choking the life out of it. Nothing was more sad on that dismal plateau than to see the struggle which was raging dumbly between host and parasite—the one immense, stern, and upright, the other insignificant, crawling, deadly. On the edge of the forest great bare masts, shrouded with this gnawing death, which hung in tattered festoons from the stumps of branches, rose grimly into the sky. Inside the forest huge trunks had come crashing down through the bamboo brake, and lay encased in moss, sleeping their last long sleep.

After skirting several forested hill-tops I returned to the open valley, some distance below the pass, picked up a trail, and wandered rather aimlessly along by the growing stream of peaty water. The trail itself was good, though I was rather surprised to see no mule tracks, and to find no trace of either animals or escort after an hour's walking, but it scarcely occurred to me yet that I was on the wrong road.

And now came an incident which for the first time set me pondering. Looking back up the shallow valley towards the pass we had crossed, I saw a Tibetan caravan coming towards me, perhaps half a mile distant, but when I looked

Plate IV

Temple and Pagoda at T'eng-yueh

On the Li-ti-p'ing; grass-land and fir forest

again a few minutes later, it had disappeared, and I never saw it again.

By this time I had begun to realise that I was not on the main road, for where was my caravan, and why had not the men halted for lunch? It was already past mid-day and I had parted from them about ten o'clock. To crown all, the path came to an end and, unwilling to retrace my steps, I decided on a new and fatal plan of action.

I have already stated that we had crossed a pass, and I had good reason to believe that streams flowing down this side reached Wei-hsi; therefore, I argued, by following one of the streams I should eventually arrive at the city, though it was only to be expected that it would take considerably longer. Why did I not retrace my steps up the valley to the point where I had parted from the caravan, and carefully follow the mule tracks? Scores of caravans use this road, and had I thought for a moment, I should never have gone wrong at all. I think it was partly from a love of plunging into anything which offered a certain amount of novelty, and partly from sheer laziness. I did not foresee any insurmountable difficulties, and though it was abundantly clear that I was some distance from the main road, I had, remembering the map, a hazy idea of rejoining it by a circuitous route without the trouble of going over the old ground again. From the hill-top where the path came to an end I turned away down the slope, and a minute later plunged into the forest in order to reach a big stream as soon as possible, where progress, I thought, might prove easier. However, as I proceeded, it became infinitely more difficult, for as the hill-tops of the plateau country were left behind, the deepening valleys became choked with dense bamboo brake, to the exclusion of everything else.

For two hours I blindly fought my way through this jungle, the bamboos reaching a height of fifteen to twenty feet and growing so thickly that I had to force the stems apart, clambering over an occasional tree-trunk and plunging knee-deep into icy torrents, while the sweat rolled off me. Sometimes I emerged momentarily from the brake, hot and angry, and finding a trail, recklessly followed it till it disappeared, but always I came again to this appalling fence of jungle, which was slowly crushing the strength

out of me. Even the beautiful sight of masses of the blue
Primula sonchifolia in the dampest parts of the forest,
sometimes growing right in the icy water derived from
the melting snow, failed to compensate me for this torture,
or to rouse my enthusiasm.

About this time a big valley, into which all these count-
less torrents poured their water, came in sight, and I
determined to try and reach it before nightfall. Though
the sun was veiled I knew my bearings roughly, having
the sense of direction well developed, and the valley in
front of me certainly lay east and west.

If the stream flowed westwards, as I supposed, I must
soon get down to the city, for by this time I had come
some distance south. But if it flowed eastwards?—and a
horrible doubt assailed me.

It had been a wearying day, but dusk was now closing
in, and with it came the rain. Darkness set in early, for
the days are short in this latitude even in summer, and the
gloom was intensified by the heavy clouds. Following a
path, I climbed one more grassy hill which promised an
extensive view from the summit, hoping that at dawn I
should be able to make certain of my bearings. But commanding as the position was, the weather showed no signs
of improvement, for the soft clinging rain clouds were now
settling down all round me. On the edge of the forest
I sat down to await the dawn and take stock of things,
wondering uneasily if there were any wolves prowling
about on such a night. My mackintosh had been ripped
to shreds while I was buffeting my way through the brake,
but I covered over my head with what remained of it; there
was also my gun and one cartridge to be reckoned with,
and this I kept handy in case of a wolf. I had no matches,
nor could I perhaps in any case have lit a fire on such a
night; food I had none, and in addition to feeling very
hungry, I was also extremely wet from wading through
torrents in order to avoid the brake.

The big valley I had seen in the evening was still some
distance away, but I had bright hopes that it would prove
the end of my troubles, and decided to push on as soon as
it was light enough to see. So I lay down to rest, burying
my head under my mackintosh and curling up like a cat to

keep out the cold, while the rain poured steadily down and formed pools beneath me.

I wondered what my men were doing and whether they had reported in the city that I was lost on the plateau. Perhaps soldiers had already started out to look for me—they could easily track me to the end of the path, but after that a regiment might scour these forests without finding any trace of me, for there was nothing to indicate which direction I had taken.

The night seemed very long, but towards morning I dozed several times, and at length when I awoke it was already dawn, a cold grey light gradually diffusing itself over the sullen-looking sky. Clouds enveloped the forested hill-tops and lay in heavy masses in the deep valleys, but the rain had turned to snow which melted as it lay on the ground. Such a cheerless daybreak offered little encouragement to me in my task.

However, I started off at once in order to get warm and shake off the numbing stiffness brought on by the past few hours of inactivity; fortune seemed to smile on me this once, since by keeping the valley in view and heading straight for it so far as the country would allow, I struck a well-worn trail almost immediately. Climbing up and down over the endless hills for nearly three hours, I at last reached a point whence I could look straight down into the big valley, and solve the problem.

The stream flowed eastwards—I was all wrong!

This unfortunate truth had indeed been gradually asserting itself for some time past—the configuration of the mountains scarcely allowed of any other interpretation, though I was loth to admit it to myself; and disgusted beyond measure as I was at the annihilation of my vain hopes, I was not altogether taken aback.

Meanwhile the path had been growing worse instead of better and presently came to an end again at the edge of the forest, a circumstance which greatly simplified matters for it left but one reasonable course of action.

I must either break away again and try a new direction in the hope of striking another path, or find my way back to the starting point before I became more involved than I was already.

Disheartening as the latter course seemed, for I had vivid recollections of my struggles in the brake, it was the only sensible thing to do, and having finally resigned myself to it, I at once started wearily back.

There were a few anemones in flower on the grassy slopes, looking very miserable in the driving snow, besides numerous rhododendrons on the edges of the forest ; and I remembered with glee that at the base of each rhododendron corolla was a big drop of honey. However, after sucking a score of flowers without obtaining much nourishment, I started eating the whole thing which, though glutinous and insipid, was not altogether nasty.

After a weary climb, for I was now beginning to feel despondent, I arrived back at my sleeping ground, and descending to the stream, prepared for the long journey up to the plateau. First I ate a meal of sorrel and any other young leaves I could find, though most of them were either hard and leathery, or soft and covered with woolly hairs ; but after drinking my fill of the clear cold water I felt very much better in spite of the fact that the rhododendron corollas had given me a violent pain in the stomach.

My plan was to follow up this stream—the biggest flowing from the direction of the plateau—to its source ; but while making every endeavour to keep it in sight, I vowed that, come what might, nothing should tempt me into the brake again.

Starting along one of the multitude of small paths I had followed on the previous day, I watched it carefully, and though it entailed much climbing up and down the ridge, I noted with satisfaction that it skirted several bamboo forests I had deliberately plunged into on the way down.

These numerous paths are used by the Lissus, who inhabit the mountains on the east or left bank of the Mekong from Wei-hsi in latitude 27° 10′ to latitude 28°, whither they have spread from their preserves on the Upper Salween. They come up to the Li-ti-p'ing in the summer to tend flocks, cut wood for their bows, collect wild honey for food, poison for their arrows, firewood, and other jungle produce, and had I followed down the streams far enough, I should doubtless have come across Lissu huts before long.

One by one the landmarks came back, and I made good progress up the valley. Only one point troubled me. On the previous day I had burst into this valley after some two hours' struggling in the brake, an experience I was not inclined to court a second time. But could I pick up a path which would lead me back to the plateau without retracing my footsteps? Nay, even if I attempted the brake again, could I retrace my footsteps?

To the latter question the answer was, obviously, no, and therefore the only course was to continue up the stream to its source, leaving out of account the tributaries I had followed through the brake. This scheme at all events promised as well as anything else. It was cruel fate that as I wandered along, laying plans to be observed after I reached my last landmark, I should flush a snow pheasant, the original source of all the trouble, and the only one I saw during my two days' sojourn on the Li-ti-p'ing. Needless to say I did not put up my gun in time to shoot it, but it occurred to me that, as there were several birds about, I might as well shoot something; and in order to make certain of my one remaining cartridge I fired point blank at an unfortunate little finch that was sitting on a bush. When I came to pick him up, I found that the No. 6 shot had not only killed him but very nearly plucked him as well, and with the exception of the feathers, entrails, and beak, I ate him entire. After a short rest I felt much better, and pushed on for the turning point, which I reached much sooner than I had expected, thanks to following a path all the way.

And now occurred a temporary check, for I could find no continuation of the path, and after wading across several streams in an endeavour to follow up the main valley, I was once more faced by bamboo brake.

Retracing my steps, I struck off into an open grass-land valley which promised an easier route—the presence of grass-land was a welcome hint of the high plateau—and plunging into more or less open fir forest higher up, I crossed a ridge, still bending my steps in what I conceived to be the direction of the pass. It was now, I thought, early afternoon.

After having crossed the ridge, I emerged from the

strip of forest into an open grassy hollow surrounded by woody hills, and there stood an old sheep-pen. I had passed that sheep-pen on the previous day just before the first path came to an end and I plunged into the brake. I was all right.

In the red mud of the path I could just distinguish my footprints of the previous day, now almost obliterated by the rain; but there were also clearly visible the footprints of two other men which certainly were not there before, probably those of soldiers who had tracked me thus far, and I shouted several times without, however, obtaining any answer.

The revulsion of feeling would have been greater but for the fact that I knew it would take me some time to reach the pass from here, and several hours to get thence to Wei-hsi. Nevertheless I set out with a light heart and a feeling as of a great load lifted off my mind, tried a short cut but lost my bearings, returned on my tracks, and at last found myself back on the main road, near the pass. There was no mistaking it this time, for the mule-tracks showed plainly enough that it swung away to the west across the valley I had so persistently followed southwards and eastwards.

A long climb over the plateau, splashing through the most appalling mud and half-melted snow, thoroughly ploughed up by mule traffic, brought me at last to a second pass, and now the forested plateau fell away abruptly below me, and tailed out in high spurs to a broad flat valley.

The rain had ceased, and to cheer me up the sun flashed out from beneath a bank of clouds for a few minutes before sinking down behind a high range just across the valley; and far below, Wei-hsi glistened in the golden light of the setting sun. It is to be noted that the first pass we crossed was that over the main watershed between the Kin-sha and the Mekong river systems; the second and higher pass, on the other hand, separates the waters flowing past Wei-hsi from those which, gathered throughout the length and breadth of the Li-ti-p'ing, flow out at Ka-ka-t'ang to join the Wei-hsi river fifteen miles below the city. Had I continued down those seemingly endless valleys, therefore, I must eventually have come out at Ka-ka-t'ang, though it is probable, as already remarked, that I should have come across Lissus before finally emerging from the mountains.

From the pass I descended the extremely precipitous road as rapidly as possible, but though the city had looked so near, it was in reality some distance away, and darkness was coming on apace, so that I soon realised the impossibility of getting in before dark.

Now I could see the lights of the city and hear the howl of a pariah dog; but in the tree-girt ravine, which gradually widened out between the towering spurs as the valley was approached, it was pitch dark.

Progress became slower; extraordinary hallucinations grew upon me, and I found myself continually halting to step carefully over large boulders which did not exist except in my imagination, while in doing so I blundered clumsily into every obstacle which the path presented, slipped over the bank on one side, and walked into the bushes on the other. Helpless birds fluttered along the ground in front of me, so that I stooped down on more than one occasion to pick one up; strange animals moved in the thickets; every light visible in the city was dancing up and down like a will-o'-the-wisp, and some poplar trees along the sky-line to the right seemed to be swaying violently to and fro as though bending before a gale, yet the night was perfectly still.

Stumbling and tripping, I moved cautiously forward with arms outstretched, weary, but in excellent spirits; stars were winking overhead, and after the cold and wet of the plateau, the air felt warm; also my clothes were dry, a fact which greatly added to my comfort.

Suddenly the dark outline of a house loomed up in front of me, and at the same moment several dogs began to bark furiously.

Entering the yard I made my way to the door and rapped loudly, but the dogs had already drawn attention to me and someone was even now approaching. Next moment the door opened, and an old woman stood before me, a flaming pine torch held high above her head as she peered into the night.

"Why it's his Excellency!" she exclaimed in astonishment, and ran back to bring her husband. My fame had preceded me!

The old man, who was half-blind, presently appeared, and taking me by the hand led me in, and so up a ladder to a large room, where in a few minutes he had a fire blazing; meanwhile his wife brought up hot water, some eggs, and a quilt, while the son, a fine strapping young fellow, set out immediately for the city, a mile away, to inform the official of my safe return and rouse my men.

How shall I describe the kindness with which those humble Lissus took me in and in the most hearty manner put everything they possessed at my disposal? However, I was too tired to eat, and bidding the old man give me one of the soft brown sugar bricks which the Chinese manufacture, I sliced off some parings and ate them; but the sweet matter burned my tongue, and for a day or two it was so badly furred that it pained me to eat anything.

My one desire now was to sleep, and the old man himself having taken off my boots and puttees—the only things I did take off—I wrapped myself in the quilt and lying down on the hard boards beside the fire soon fell into a deep dreamless slumber.

About nine o'clock I awoke suddenly to find the room full of light and noise. However it was only Kin and Sung who had arrived from camp with food and blankets for me, but I was too weary to move, and bidding them put out the torches after covering me up with extra blankets, I turned over on my hard bed and immediately fell fast asleep again.

On the following morning I awoke early, and after a hearty breakfast rode out to where the men had pitched the tents, on the edge of some rice fields just outside but overlooking the city. It was a glorious morning, delightfully warm, and I felt in the best of spirits.

My Lissu friends at first refused to take anything for their hospitality, but later I sent them a present of money which they gratefully accepted; and during the day, as I had anticipated, there arrived back several soldiers who had been sent out by the official to look for me—a kindly act of grace on his part. The soldiers told me that they had followed my tracks to the end of the path, and had then been at a loss which way to turn; for there was nothing further to indicate my route.

The official himself sent round a polite message to the effect that he would be pleased to see me, and on the following day I went round to the Yamen accompanied by Ho-shing. On the way I made him coach me in the things I ought to say, and the answers I might expect to receive, though Ho-shing was a difficult person to understand, on account of his vile patois.

The T'ing, a middle-aged, ascetic-looking, but rather handsome man, somewhat reserved and cold in manner, greeted me with every mark of politeness, and smiled drily for the first time when I thanked him for sending out soldiers to look for me. He made no trouble about my going on to A-tun-tsi, and indeed wrote out directions to his subordinates that I was to be supplied with an escort of three soldiers and receive every facility on the way, but like all Chinamen, he exaggerated the difficulties of the road ridiculously.

As a parting gift he sent me round a haunch of bacon, a chicken, and some eggs, and not to be outdone in generosity, I retaliated with two cakes of scented Vinolia soap, which I could certainly ill afford. Taking it all round the T'ing and I became quite friendly, a circumstance which later was to stand me in good stead.

CHAPTER IV

UP THE MEKONG VALLEY

WE rested four days at Wei-hsi. It is a small unwalled city with cobbled streets, patches of cultivated ground being freely interspersed with tiled houses of wood or mud bricks; but the pear trees now in full blossom gave a brighter appearance to the shabby temples and mean lanes.

The valley itself is little more than a mile across here, considerably less elsewhere, shut in by the mountains of the Li-ti-p'ing to the east and by a high ridge to the west, clothed below with scrub and pine trees harbouring two or three species of pheasant, and above with fir forests. There is no flat ground anywhere except immediately beside the river, where a small flood-plain dotted with bushes affords opportunities of shooting duck and snipe. Though flowing with a strong current, the stream is of small size, fordable almost anywhere at this season, and spanned by a single wooden bridge of no interest. The gentler slopes of the valley are extensively terraced, rice being the principal crop for the whole of this district. Even the city is built on a slope. Major Davies gives its altitude as 8000 feet, and my aneroid readings agreed very closely with this.

Considering that we were nearing the end of April, the vegetation was not far advanced, certainly not more so than it is in England by the end of the first week of that month, and there was snow on the ridge immediately to the west, probably not more than 2000 feet above the city. The valley itself however is dry, like the Mekong valley below, and the most interesting plant I met with during several

Plate V

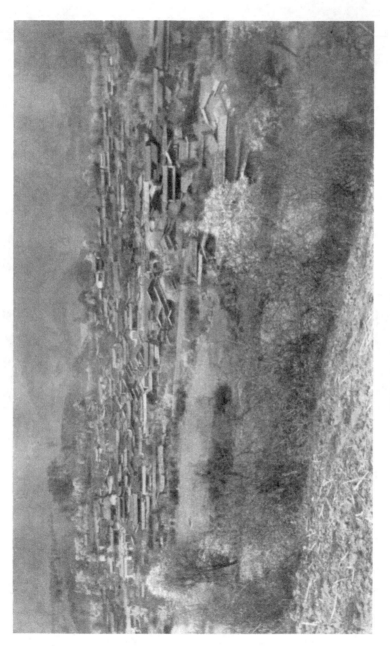

Wei-hsi-t'ing, seen from camp

rambles was a crimson-purple Primula (*P. vittata*) coming into flower in the marshes.

Like all frontier cities, Wei-hsi is a great trading centre, and a motley crowd, amongst whom may be seen sallow-faced Chinamen, fair-skinned Mosos, swarthy Tibetans, and sullen-looking Lissus, throng the main street on a market day.

The surrounding country is inhabited chiefly by the Moso tribe, whose centre of distribution is the strip of land enclosed by the great loop of the Yang-tze already referred to, though they extend as far west as Wei-hsi and up the Mekong valley as far as Tsu-kou. They are most numerous in the neighbourhood of their ancient capital, now a Chinese prefectural city, Lichiang-fu.

I can imagine nothing more charming, nothing in better taste than a well-dressed Moso girl. She wears a white, or perhaps dark-blue skirt, closely pleated lengthways after the manner of a skirt-dancer's costume, reaching well below the knees; a dark blue blouse tied round the waist; and a head-cloth of a dull red colour, above which is bound the queue, probably adorned with jewels of coarse workmanship, silver set with coral and turquoise, to match the long pendent ear-rings. Nor is it too much to say that many a Moso girl is wonderfully pretty, with a round good-natured face, regular features, a light complexion which is most readily described as sunburnt, and large dreamy eyes, though the general expression is one of considerable animation.

Frequently they bind their ankles with narrow strips of tape, a custom to which it is difficult to assign any reason as it stands. I eventually came to the conclusion that it might be the relic of an ancestral custom, comparable to those rudimentary structures which, of no use now, yet survive in many animals, a sore puzzle to the anatomist, who sees in them a remnant of some previous organ which through disuse has dwindled to a mere cipher of its former complexity, though still perhaps leaving a clue to its history. But if so, what can these ankle bandages, of no practical utility, though ornamental, represent? Probably ankle-bracelets, such as the Burmans and other peoples from this part of Asia wear to this day. A metal hoop round

the ankle would obviously be inconvenient to an active
people like the Moso, who cultivate steep hillsides, and
carry heavy loads a long distance to market—always a
woman's job.

The women, as remarked above, adhere loyally to the
tribal customs, particularly in the matter of attire, but the
men, as is so often the case, have adopted Chinese dress,
no doubt largely as a measure of convenience ; and all the
Mosos one meets with, along the main road at any rate,
speak Chinese better than most other tribesmen who profess
to do so.   This is the case both with men and women,
though usually while the men of a tribe can speak Chinese
more or less, their women folk cannot.   This is particularly
true of the Tibetans in many places.

The Mosos are a medium-sized race, the women being
bigger and more healthy-looking than Chinese women; but
though sturdy of limb—on the return journey I had several
Moso girls as porters, carrying their loads by means of a
strap passing round the forehead—their figures were nevertheless trim.   They may be of Tibetan origin, as is generally
accepted, though the fact is by no means obvious from a
superficial acquaintance.   Personally I believe them to
have come across the mountains from the west, and to be
much more closely allied to the Lutzu and Lissu tribes than
to the Tibetans ; they appear to me to have attained high-
water mark in the evolution of an emigrating jungle tribe,
though it is well known they have been in their present
home for several centuries.   On the other hand they possess
a written language akin I believe to Tibetan, and to some
extent, particularly in the matter of dress and jewellery,
resemble the Tibetans of the Pang-tsi-la region, though
this may be simply owing to contact in the Yang-tze valley,
unless indeed the so-called Tibetans of Pang-tsi-la are
really Mosos.

But free and unrestrained in their manners as are the
Mosos, taking photographs of them is quite another thing,
and they are not easily persuaded to stand up to the
camera, at least not in the city under the scoffing eye of
the Chinaman.   I believe the little minxes know they are
good-looking and realised that I was trying to put the fact
on record, for with laughing eyes they sedulously avoided

me when I went into the city with my camera, while the children positively ran. Not so the Tibetans however, reckless fellows; they would always face the camera for the sheer adventure of the thing, nudging each other and screaming with laughter, and for a few cash they would even dance to the accompaniment of their squeaky bamboo fiddles.

We left Wei-hsi on April 22, and marching down stream, camped for the night below Ka-ka-t'ang. The hill sides became still more dry and bare as we descended, a good deal of purple shale being exposed. Small bushy oaks, very prone to attack by a curious mistletoe (*Viscum articulatum*), with pines and rhododendrons, made up the bulk of the flora, and the ride was uninteresting, though the small arched bridges of stone, pointed rather than rounded, which spanned the tributary torrents such as the Ka-ka-t'ang stream, were quaint.

Continuing next day down a narrow path high above the torrent, we caught sight of the Mekong-Salween watershed, still covered with winter snow, framed in the mouth of the gorge, and shortly afterwards we descended to the Mekong itself.

Hsiao-wei-hsi, reached the same evening, marks practically the southern limit of the rainy belt on the Upper Mekong, south of which we enter upon the dry region, distinguished in turn from the true arid region to the north chiefly by the presence of the 'spear grass' (*Stipa* sp.) and a few shrubs. But of these three regions into which the Upper Mekong naturally divides itself owing to the peculiar distribution of its rainfall, we must say more later.

The river here flows' through a comparatively broad valley flanked by double and triple ranges of mountains, showing signs of a heavier rainfall than prevails to north and south, where the gorges shut in the river again. On the left bank, up which lies the road, cultivation is extensively carried on, rice being by far the most valuable crop; and all the great alluvial fans which have been washed down by the torrents on this side are skilfully terraced. Wheat, tobacco, cotton, beans, peas, buckwheat, 'red pepper' (*Capsicum*), and several kinds of vegetable are all grown in lesser quantity; and hedges of pomegranate, with walnut, orange, persimmon, and pear trees, are scattered down the slopes or give shade in the villages.

I stopped a day at Hsiao-wei-hsi in order to climb the mountain and examine the flora of one of the deep shady gullies which, in comparison with the dry exposed slopes where there is little else besides scrub oak, pines, rhododendrons and a few other Ericaceae, present a rich assortment of trees, ferns, and rock plants. In the gully I found *Primula sinensis, Saxifraga sarmentosa,* an *Epimedium,* several *Crassulas* and numerous orchids, not however in flower, with such woody plants as *Hydrangea, Philadelphus* ('syringa'), *Decaisnea,* numerous Caprifoliaceae, Celastraceae, and so on—a totally different and far more varied flora.

A little below Hsiao-wei-hsi we saw for the first time the rope bridges about which we had already heard so much; but as this structure does not attain to its most ingenious form till we reach Tsu-kou, consisting here only of a single two-way rope, instead of two one-way ropes, I will postpone a description of it for the present.

Soon after we reached K'ang-p'u on the second evening, the Tussu, hearing that I was interested in flowers, sent me round some magnificent pink paeonies, sweetly scented, and a big spray of some orchid—a *Dendrobium* I think, and later I took a cup of tea with him, and was given some of the juiciest oranges I have ever tasted, besides excellent pomegranates. In the garden behind the little yamen were bushes of pink and yellow roses, paeonies, and orchids in large earthenware pots formally arranged down the path, and beyond, a wilderness of weeds and cabbages. Above the gently-sloping rice-fields, brooding over all, were the dark forested mountains crested right along with winter snow.

The weather had now set in fine, and nothing could have been more delightful than these marches up the Mekong valley, for we took matters fairly easily, making four stages from Hsiao-wei-hsi to Tsu-kou. Sometimes the narrow path was enveloped in the shade of flowering shrubs and walnut trees, the branches breasting us as we rode, the air sweetened by the scent of roses which swept in cascades of yellow flowers over the summits of trees

Plate VI

Descent from Wei-hsi to the Mekong Valley;
Mekong-Salween Divide beyond

thirty feet high; sometimes we plunged into a deep limestone gorge, its cliffs festooned with ferns and orchids, our
caravan climbing up by rough stone steps which zigzagged
backwards and forwards till we were out of ear-shot of the
rapids in the river below; sometimes the path was broken
altogether by a scree-shoot, which, dangerous as it looked,
the mules walked across very calmly, though sending rocks
grinding and sliding down through the trees into the river.

In one gorge through which we passed, large pot-holes
were visible across the river between winter and summer
water marks and yet others still higher up, forming a conspicuous feature of the otherwise smooth bare cliffs which
dipped sheer into the river; but on the left or shaded bank
dense vegetation prevailed wherever tree, shrub, or rock-
plant could secure a foothold. The further north we went,
the more rich and varied became the vegetation of the
rainy belt, though the paucity of forest trees, except deep
down in the gullies, was always conspicuous.

Shales and slates, dipping at very high angles, and
often vertical, alternated with limestone, through which
the river had cut its way straight downwards; but at one
spot, where an enormous rapid had been formed, huge
boulders of a dark green volcanic rock, like lava, with
large included fragments, lined the shore and were piled in
confusion below cliffs of slate.

It is at sunset that the charm of this wonderful valley
is displayed at its best, for the sun having dropped out of
sight behind the western range still sends shafts of coloured
light pulsing down the valley, rose, turquoise, and pale
green slowly chasing each other across the sky till darkness
sets in and the stars sparkle gloriously. It is long after
dawn when the sleeping valley wakens to floods of sunlight
again, and the peaks which stand sentinel over it, blotting
out the views to north and south, lose the ghastly grey
pallor of dimly-lit snow.

On the third day I noticed that one of the muleteers, a
Lissu from Hsiao-wei-hsi, was limping badly, and in the
evening I looked at his foot, which had been severely cut
on a rock; for most of these men go about bare-footed.
The sole was protected by a layer of tough horny skin
three-eighths of an inch thick, grimed with dirt, and so

hard that I had great difficulty in cutting through it to open up and wash out the wound. After putting on a little iodoform, I bandaged it and next morning sent the man home, for indeed he could scarcely walk. The sequel to this, the first appearance of my medicine chest, was that all the muleteers discovered that there was something wrong with them—not because they wanted to go back, but because they wanted to sample my tabloids. Even Ho-shing joined in the popular demonstration, first asking me to cure his friend (procured from the village at short notice) of asthma, and then informing me dejectedly that he himself suffered from running at the nose. This last was a disorder that nonplussed me, so to uphold my fledgling reputation I laughed it off and told him to go away.

Passing through several fine gorges lined with tall juniper trees we arrived opposite Tsu-kou about five o'clock in the evening of the fourth day and continued a mile up the river bank to the next rope bridge, just below the larger village of Tsu-chung, also across the river, and now we had our first experience of the single-way rope bridge.

The rope bridge is not properly a bridge at all, though it is a rope, made of twisted bamboo strands, its diameter being barely three-quarters of an inch, so that a single rope, perhaps fifty yards in length, weighs only about twenty pounds. Like many other things which are ingenious, the principle of the rope bridge—for such I must continue to call it—is delightfully simple. The rope is twisted several times round an upright wooden post fixed high up on one side of the river bank, a small platform being cut out to serve as a landing stage, and stretches across the river to another post fixed lower down on the other bank, where there is also a landing platform. Thus it slopes steeply from one bank to the other, though the slope is as a matter of fact nothing like so steep as it appears, the difference of level between the two posts probably never exceeding twenty feet, while the river may be thirty yards across.

On the top of the rope runs a slider, which consists of rather more than a half-cylinder of tough wood nine inches to a foot long, and between one and two inches in diameter, polished smooth inside and well greased before use. Slots

Plate VII

The Mekong at Tsu-kou

A Rapid on the Mekong, below Tsu-kou

are cut transversely through the convex upper face of the semi-cylinder, near each end, and beneath the handles so blocked out are threaded leather thongs.

If the gradient of the rope is small, so also is the friction of the slider, and the weight of the rope causing it to sag towards the far end, the rope dips more rapidly to begin with than it would if it were rigid; consequently the impetus gained at the start is sufficient to carry one across the river. On the other hand the sag, still further emphasized by a person's weight, causes the extreme end of the rope to slope up towards the landing stage, so that the person crossing usually has to pull himself up a few yards at the end.

It will readily be seen that, since this is a single-way rope bridge, there must necessarily be two of them at each crossing, one each way; moreover, that the method is only practicable where the banks are steep and the river comparatively narrow. Of the dozens of single-way rope bridges I saw, I do not suppose one of them spans more than thirty yards of water. For this reason they are not found on the Yang-tze, though they occur a long way east of it, on the Yalung river.

On the other hand, given a narrow river and steep cliffs, there is no limit to the height at which the rope may be suspended above the river, and the lowest point varies from a few feet above the water at winter level—such ropes being partly under water during the summer rise, and therefore impassable—to as much as eighty or a hundred feet above, as is the case with the rope at Samba-dhuka, also over the Mekong, which spans an extraordinary gorge measuring about twenty yards from cliff to cliff. Alarming as these greatly elevated ropes appear, however, it would really make very little difference whether one fell into the Mekong from a height of one hundred feet or a height of five feet.

As far as the traveller's own feelings are concerned the crossing of the bridges seems to be accomplished with the speed of an express train, but Mr Edgar, of Batang, who carefully timed several journeys, obtained an average speed of only ten miles an hour, though animals may possibly travel twice as fast. Of course different ropes vary, but the speed is nothing like as great as one is tempted to believe.

But to return to our own experiences. I had sent two of the soldiers on ahead to procure slings from the nearest house, and when I arrived at the bridge I found them already on the scene. The soldiers who had escorted me from Wei-hsi, however, were all afraid of the rope bridge now they found themselves face to face with it, and Ho-shing was on the verge of tears at the prospect of crossing. Seeing that we should never get over at this rate, I decided to go first myself, though as a matter of fact I felt as diffident as anybody; and my feelings were still further harrowed by the performance of a soldier, who, after allowing the Tibetans to tie him up, thought better of it just as they were about to let go, and with pitiful entreaties begged to be untied.

I now stepped forward, and with as little delay as possible the Tibetans suspended me skilfully from the well-greased slider, one pair of thongs passing under my thighs, the other under my arm-pits, so that I hung close beneath the rope; and thus secured I advanced to the edge of the platform. Everybody crowded round and gave me excellent advice which I did not understand, telling me exactly how it was done; but as it was obvious that the only point of importance was to keep well clear of the rope, while for the sake of comfort I kept a firm hand on the top of the slider, I did not pay much attention to them.

" Let go!" and at the word I was whirled into space. Whiz! a rush of air, a catch of the breath, a smell of something burning—the rope gets very hot—the hum of the slider over the twisted strands, a snap-view of the muddy river foaming below, and I was slowing down where the rope sagged at the other end. It was all over in a moment, and pulling myself up the few remaining feet to the platform, I untied and stood up on the opposite bank. After that first experience there was nothing I enjoyed so much as a trip across a rope bridge.

The boxes were slung over in exactly the same way, but as they were not capable of hauling themselves up to the platform, it was sometimes necessary for a man to attach himself to a sling, work his way slowly along the rope, and twining his legs round the load, pull it up after him. After landing I climbed up the bank and began to look about for a convenient camping ground, for it was growing dark and threatening rain. Just then a Tibetan came along and seeing me he stopped, smiling in friendly welcome, and asked me where I was going. I told him that I had just arrived from lower down the river and intended to camp at Tsu-kou for some days at any rate, whereupon he became full of excellent suggestions, pointed out a good camping ground in a gully close at hand, and went off to his house a few hundred yards distant to get a mule.

The camping ground which I had adopted at his suggestion was a small level corner surrounded by trees and boulders, near the bottom of the deep gully referred to, down which roared a big torrent from the western range. Here was plenty of room for the two tents, and a large block of stone near by afforded a deep recess in which boxes might be stored out of the rain. I therefore descended again to the landing stage, and finding that some of the boxes had arrived, Kin and I started carrying things up the cliff to the selected spot; it was just dark now and rain had begun to fall heavily, so that we hastily bundled everything, including a supply of firewood, under the rock. Presently my Tibetan friend arrived with his mule, and the remaining things were soon brought up from below, all the men helping to erect the tents, light the fires, set the water boiling and do odd jobs round the camp under the glare of pine torches supplied by the Tibetans. By this time the rain had ceased and presently the stars were shining brightly, and after supper, weary with travel and hard work, I turned in and was quickly lulled to sleep by the roar of the torrent.

Next morning I was awake early, and went out to look about me while breakfast was being prepared. The small flat on which we were encamped was well screened by large stranded boulders which had been rolled down by the torrent, and was protected by a high bluff. Above the torrent the gully was densely clothed with trees and shrubs, Viburnums, honeysuckles, *Diervilla*, shrubby euphorbias, Celastraceae, barberry, the beautiful *Chionanthus retusus* with its dense masses of white flowers, magnificent walnut trees, *Pyrus*, maple, alder, and many more; in strong contrast to the exposed hill sides above. Higher up were woody climbers, a beautiful clematis, the crimson-flowered *Schizandra chinensis*, *Aristolochia monpinense*, with flowers very similar to the familiar Dutchman's Pipe (*A. sipho*), the curious *Akebia*, *Sabia yunnanensis* and other delightful things, with splendid trusses of white rhododendron and numerous orchids, including *Cypripedium luteum*.

The dry hill slopes at this low altitude, however, afforded little of interest, the familiar yellow mullein (*Verbascum*), a handsome purple-flowered broom-rape (*Orobanche*), and a boraginaceous flower of brilliant hue (*Cynoglossum amabile*) being the most noticeable. Some 1500 feet above the Mekong there was hidden away a splendid patch of white opium poppy, secure from the vulgar gaze; it covered 75 square yards, and was now in full bloom.

To the west the river was shut in by a double range of mountains, the lower, which alone was visible from below, rising about 3000 feet above the river and cresting the steep slope with high limestone towers and cliffs; and as far as I could see the same physical features occurred on the east or left bank. Between the river and the mountain slope extended a narrow platform along which a few huts were scattered from Tsu-kou to Tsu-chung. Below Tsu-chung rice is cultivated, this being the last village on the Mekong where it is known, for immediately to the north the climate changes abruptly. With the climate and the want of rice the people change also; there are no more Mosos or Lissus now, only Tibetans, and such Chinese merchants as have settled at the big trading centres.

The platform just referred to however, which is only a few hundred yards wide, does not represent all the cultivation, for the slopes of the first hill range have to a large extent been cleared of their open pine woods and rhododendron scrub, and here are raised crops of wheat and barley. It is wonderful to see the Tibetans ploughing these hill sides with bullocks drawing the wooden plough, which is so light that the man picks it up at the end of the day's work and walks home with it. So steep are the slopes that the stones come rattling down the hill side as each furrow is turned, and so poor the soil, choked with

Plate VIII

The Mekong at Tsu-kou, looking north; Mekong-Yang-tze
Divide in the distance

big slabs of shale, that the very thinnest of wheat crops comes up, and a much larger area is under cultivation than would seem necessary from the size of the population.

Tsu-kou itself boasts scarcely a dozen huts, and scattered up the cultivated slopes of the barrier range are other small villages. Before 1905 the Catholic mission was established in Tsu-kou, but after the burning and sacking of the village by the Tibetans, who rose against the Chinese all along the border from Batang southwards, it was rebuilt at Tsu-chung which is now the more important village, boasting perhaps a score of families, besides a garrison of thirty Chinese soldiers. The charred walls of the mission house at Tsu-kou still stand to mark the site, and close by are the graves of the two French priests who perished in the disaster.

The barrier range above the river has been blocked out from the main range by the torrents which flow diagonally towards the Mekong in the upper part of their course, thus cutting out a wall of rock in front of them, through which they are eventually compelled to saw a deep gorge to the river, though a few of them plunge over the precipice.

The tortuous courses of these streams, with the consequent formation of barrier ranges, seems to be largely due to an unequal distribution of rainfall in the mountains and in the valley aided by the general N. and S. dip of the strata. Here there is a very heavy rainfall on the Mekong-Salween watershed throughout the summer, and a moderate one in the valley; but as we go further north there is an equally heavy rainfall on the divide, while only a negligible quantity of rain falls in the semi-desert valley; and immense spurs buttress the watershed from river to crest, with no sign of any barrier range. The same structures occur on the Salween on a still larger scale in the rainy and arid regions respectively.

After four days in camp with my three men, the local Tussu requested us to move into one or other of the villages, as he was afraid that we might be plundered in the night by mountain robbers. Meanwhile I had made friends with the two Catholic priests at Tsu-chung, Pères Mombeig and Lesguergues, who visited me in camp and several times invited me to take supper with them, showing me over the cathedral—a wonderful little building considering the limited resources of the district—and helping me in many ways. Père Mombeig had been through the rebellion of 1905, and from the wild mountains to which he fled from his pursuers, had watched his home and church burning. Then for many days he had wandered half-starved in the mountains and amongst the Lutzus and Lissus of the Salween valley, before finally escaping to the south.

It was through Père Mombeig also that I secured the services of Gan-ton, the man I had met when I first crossed the river, as guide and interpreter during some of my journeys, though I cannot record that he turned out entirely satisfactory. However, as he plays an important part in my first two journeys, I must say something about him.

He was in the first place that rather curious mixture, a Catholic Tibetan, speaking Yunnanese and Lutzu fluently, and sufficient of the Moso and Lissu tongues to make facetious remarks to almost any tribesman we met; but he did not take his religion very seriously, and was a proselyte or a staunch Lamaist indifferently according to the nature of the task he was called upon to perform. Nevertheless he was intelligent and resourceful, always cheerful, and though in some respects a knave, I found him invaluable as guide, interpreter, and companion during two trying journeys.

The French priests now put at my disposal a small outhouse belonging to a family of Catholic Tibetans at Tsu-kou, my men occupying another room in the same house. The houses of the poorer Tibetan families in the Mekong valley, as of the Lutzu in the Salween valley, consist usually of a single room, sometimes with a partition, very different from the large two-storied Tibetan manor houses met with elsewhere, and when other rooms are required—cattle byres, store-houses for grain and so on—they are built separately.

We had not been here long before I had to dismiss Ho-shing. For some time I had noticed that my stores were rapidly diminishing, and enquiry elicited the fact that Ho-shing, who alone had access to them, was the responsible person. Moreover Kin and Sung, who were both excellent fellows, did not get on with him at all, and personally I never took to him, so that he left unwept, Sung taking his place as cook, while Kin accompanied me on my expeditions up the mountains.

During the twelve days we were in this neighbourhood I explored the gullies and barrier range without much success. It was impossible to get far up the streams, where many interesting plants occurred, on account of precipices and gorges, while on the other hand the exposed slopes offered little of novelty. One day however Gan-ton led us over the barrier range, and so down into the valley beyond, above the fall; and at once we were in a new world. From the pass (10,000 ft.) we looked down over dense forests of rhododendrons, great trees thirty feet high, many of them in full bloom; higher up more forests of fir stretched out dark tongues towards the pyramidal peaks still white with snow, and to the east we could see across the Mekong valley to the snow-covered mountains of the Mekong-Yang-tze watershed.

Descending a steep path, we found a rich undergrowth coming up, but there was as yet scarcely anything in flower, though the place promised well, and eventually reaching a clearing where a small hut had been erected close beside the torrent, we halted for lunch.

By following up this valley Gan-ton informed us, we could cross the Mekong-Salween watershed, and reach Tsam-p'u-t'ong on the latter river, and I straightway decided to make the journey as soon as the pass was open, about the first week in June; for it was evident that here right before us was a wonderful wealth of alpine flowers.

On the way down we stopped at a mountain hut for refreshments, a simple one-roomed shanty, and I watched the men make their tea, though personally I drank yak milk.

The water was boiled in a big iron pot suspended over the fire, and a handful of coarse tea, sliced off a round brick, thrown in. After being well stirred, the liquid was poured through a conical basket-work funnel into a tall wooden cylinder bound with bamboo hoops; in the funnel had been placed a lump of butter which, melting and passing through the tea leaves with the hot tea, now swam on the surface in large oily drops. The basket was then removed with the tea leaves, a pinch of coarse salt thrown in, and the mixture stamped vigorously up and down in the cylinder by means of a perforated wooden disc attached to a long handle, which fitted closely like a piston. The oily drops of butter are thus thoroughly broken up and emulsified, the salty flavour equally distributed, and the beverage made ready for consumption. Taken hot from a cup, as tea, the Englishman is apt to find it nauseating, particularly when there are yak hairs from the butter generously distributed through it; but taken hot with a spoon, as soup, it is quite palatable. Such is the power of the association of ideas.

But the Tibetan himself does not as a rule drink it. He takes from the ample folds of his cloak a small leather bag of *tsamba* (roasted barley ground into flour) and a wooden bowl—two of the numerous articles a Tibetan always carries about with him in lieu of much superfluous clothing—and mixing a little *tsamba* with the tea, he kneads the mass into a ball of dough-like consistency, and complacently chews it, powdering white rings round his mouth in the process. Then out come the long pipes, a dry tobacco leaf is plucked from a bunch stowed away in the corner, crunched up in the hand, and the dust dropped into the pipe bowl, which is lighted with a layer of hot charcoal ashes pressed down on the top; and the men sitting cross-legged round the fire in the middle of the room contentedly smoke.

Just before we left Tsu-kou, an unfortunate incident occurred. My landlord brought his youngest child to me one evening and asked for medicine for him. The little fellow, who was only about a year old, was suffering from fever and a severe cough, having evidently caught a chill; and to ease the coughing I at once gave him a few drops of chlorodyne. Then giving the parents some quinine tabloids, and telling them to wrap him up carefully, keep him in a warm room, and feed him lightly on hot milk, I dismissed the affair from my mind.

On the following afternoon it seems, the parents, who had told me in the morning that the child was better, suddenly anticipated the worst, and very foolishly carried the sick child to Tsu-chung to be baptized! Poor ignorant people, they were more anxious to save his soul than to preserve his body!

I was at supper in my room that same evening when suddenly I heard the wail of a woman, and next minute Kin came in to tell me that the baby had just died quite peacefully in his mother's arms. I went out to see if anything could be done, but as I could find no sign of breathing with a looking-glass held over the mouth, I concluded that the parents were right; the little baptized soul had fled.

The mother, a Tibetan woman, warm-hearted and affectionate as are all Tibetan women by nature, was greatly distressed, but her more stolid Chinese husband showed no visible emotion; three small children stood silently by, with their fingers in their mouths, and their little black eyes wide open. Perhaps they had just seen death for the first time in their lives, and I wondered what thoughts were passing in their minds. Next morning I was awakened early by the noise of hammering, and outside I found the father busily employed. He was putting together the coffin for his little child, who was quietly buried on the hill side beneath the white rhododendron flowers the same afternoon.

Gan-ton had undertaken to find porters and make all arrangements for our journey to the Salween a fortnight hence; but unfortunately he did not keep the plans to himself. The news got round amongst the soldiers at Tsu-chung, and a few days later a runner appeared from A-tun-tsi with a note from the official there, asking me not to go to the Salween, as the Lutzu were a very wild tribe and the Chinese could not guarantee my safety. Under the circumstances it seemed best to start at once for A-tun-tsi and set at rest the suspicions of this mandarin, for I did not doubt that he suspected me of being a political agent, and had for that reason described the Lutzu, a most inoffensive people according to Prince Henri d'Orléans, in unnecessarily harsh terms; as is the way with Chinese officials, who in the matter of diplomacy, prefer devious to direct action. Accordingly we packed up, collected mules through Gan-ton who was to await my return, and on May 11 set out for A-tun-tsi.

CHAPTER V

A-TUN-TSI

It was now necessary to take the Tsu-kou mules across the Mekong, but the Tibetans made no difficulty about this, though the passage of the river occupied nearly three hours.

First the rope was tightened, for it is imperative that the mules should slide right across and land on the opposite bank, they being quite helpless until they feel ground beneath them again. The rope of course, after being used for a time, begins to get slack and sag as already described, and the great weight of the animals naturally causes it to sag a good deal further, so that unless it is braced up as taut as it will go, there is every chance of a mule sticking when little more than half-way across.

Several men therefore twisted leather thongs round the rope while another man slackened the coils round the post; and at a given signal they all heaved together, the slack so pulled up being quickly slipped round the post and the loose end made fast. The rope now stretched practically in a straight line from bank to bank.

Next, some of the men went across, slapping lumps of grease on to the rope at both ends, and as quickly as possible the boxes were slung over the river and safely landed.

Now came the turn of the animals. Each was led on to the platform beneath the rope and slung from a single slider by two loops, one passing under the belly the other under the chest, the thongs being fastened together across the back of the animal and secured by a stick thrust through the knot; meanwhile two men held back the slider to prevent it starting prematurely which it was now quite prepared to do.

Plate IX

A Mule being slung across the rope
bridge over the Mekong at Tsu-kou

A Lutzu crossing the Salween in
the jungle region

At a signal from the men waiting on the other side the animal was pushed over the edge of the platform and, clawing desperately at the ground as he felt himself slipping into space, shot off in a cloud of dust. It was comic to see the poor animals dangling limply over the river, kicking ineffectually at nothing. But all went safely across to the other side, and as they felt solid ground within reach again, they clawed desperately at the steep bank once more, two men standing ready with leather thongs twisted round the rope to stop the slider and untie the animals—a dangerous game sometimes, for the frightened brutes lashed out indiscriminately till their feet were firmly planted on the bank once more.

Slinging a number of animals across these rope bridges wears them hard, and Gan-ton told me that the Tsu-kou ropes are renewed three times a year. If we allow that on the average five men cross every day, then in round numbers the passage of six hundred men represents the life of a rope, though I doubt very much if it is really as much as this. On the other hand a score of men will change the ropes in the course of a morning, and since a pair of them cost only Taels 2·50 (seven shillings), each village paying for the upkeep of its own ropes, it does not represent a very serious outlay. When the big Lama caravans from Chinese Tibet visit Lhasa, they often carry their own rope bridge with them, a necessary precaution in view of the fact that these caravans sometimes comprise more than a hundred animals and twenty or thirty men.

On the Mekong the Tibetans have found another use for the rope bridge, namely as a means of measuring time. The day is divided into four periods, the first period while the sun is below the eastern ridge, the second from the time it tops the ridge till mid-day, the third from mid-day till it tops the western ridge, and finally from the time the sun disappears below the western ridge till dark. The second and third periods are further subdivided according to the apparent height of the sun between the zenith and the eastern or western ridges, this distance being described as so many lengths of the rope bridge, and, rough as the method sounds, it is surprising how accurately a Tibetan who understands European methods will tell the time by glancing at the sun. Gan-ton could tell the right time within half-an-hour at any period of the day if the sky was not completely overcast.

As already remarked, below Hsaio-wei-hsi, where the river broadens out considerably and the banks usually slope more gently, there are only two-way rope bridges, so that the passenger having slid down to the middle of the rope, is compelled to pull himself up the remainder of the distance.

Unlike the Chinese, the Tibetans do not use pack-racks, and though the custom of simply fastening the loads on either side of the saddle by means of a loop has its disadvantages, it certainly saves a great deal of delay when, as in the case of crossing a river, every box has to be separately taken off and tied on again. True, things get badly shaken up by this method, and occasionally a box falls off, though by no means so often as one would expect after seeing the method of tying them on ; but on the whole I am inclined to think the Chinese exaggerate the importance of pack-racks, with all the elaborate knots employed.

Bad as is the cliff road from Hsiao-wei-hsi to Tsu-kou, it becomes positively perilous beyond the latter village, more particularly where it traverses the gorges, winding sharply this way and that, ascending and descending by steep stone stairways, jumping scree shoots, or crossing a ravine by a bridge of tree-trunks. The soldiers who loafed along behind were worse than useless, for presently from far down one of the gorges would echo the melodious tinkle of bells, and in the worst possible place the leading animals would meet ; whereupon the soldiers rushing up too late only made the confusion worse.

Mules are not only stubborn, but stupid, and when unattended they will sometimes try and pass each other on a two-foot path. The loads of course lock, but the animals, instead of backing out quietly, put down their heads and push as hard as they can to try and disentangle themselves, so that not only are the boxes banged to pieces against each other and against the cliff, but the outside mule often stands a good chance of going over the precipice.

North of Tsu-kou are Tibetans only. Prayer-flags

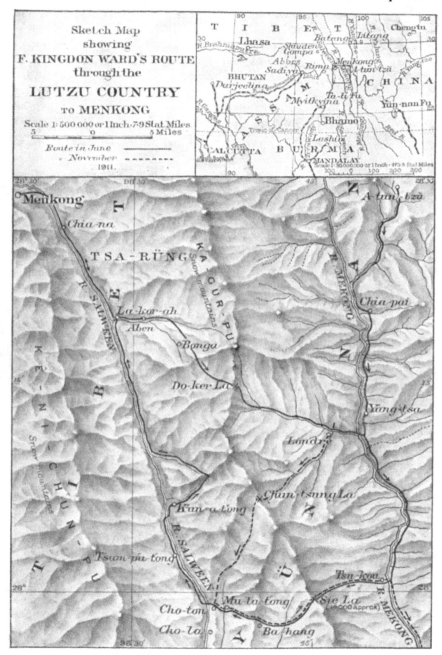

Map to illustrate journey from A-tun-tsi to the Salween and back.

[Scale: 4 miles to the inch. *By courtesy of the Royal Geographical Society.*]

flutter gaily from tall poles in all the villages, and at Yang-tsa, where there is a small temple on the opposite side of the river, half a dozen bamboo ropes are stretched across and similarly decorated.

The river here is lined with big juniper trees, and in the gorges below, where the rainfall is considerable and the vegetation correspondingly dense, the Tree of Heaven (*Ailanthus glandulosa*) is frequently conspicuous. Beyond Yangtsa however, the climate changes in a remarkable manner, and it is to be noted that this change of climate is closely correlated with the change of race already alluded to. A similar double change occurs in the Salween valley, as we shall point out presently.

The native of S.E. Tibet or Kham is an extremely tall man, averaging little under six feet, though he looks even taller by reason of his slimness—unless indeed it is his height which makes him look slim for, when stripped, the great depth of his muscular chest and the set of his powerful shoulders give an indication of his unusual strength.

I found them pleasant and friendly people, though filthy. It was rare indeed to meet a man who did not salute me by spreading out both hands in front of him, palms upwards, perhaps a survival from some form of greeting indicating that the hands conceal no weapon. Less frequently they greeted me by putting out their tongues, but when they asked for anything they always closed their fists and stuck up their thumbs. Consequently I came to regard the outspread hands as an ordinary road-side greeting, the putting out of the tongue as a more humble mode of address, and the sticking up of thumbs as a sign that a favour was being asked.

The men wear a single long robe like a dressing-gown made of sackcloth, and leather-soled cloth boots reaching to the knee; in the daytime this cloak is tied up round the waist clear of the knees, forming a short skirt or kilt, and one shoulder is usually slipped out, or in the summer time both shoulders, the sleeves then being tied round the waist. At night the Tibetan wraps himself up in this long cloak and lies down to sleep whether it be under the stars or in a house. Year in and year out, night and day, this coarse hempen cloak serves the poor Tibetan for clothing.

Inside the slack of this useful garment, when tied up during the day, a man will carry a variety of things, and it is like a grotesque conjuring trick to see him producing, apparently from the recesses of an enlarged paunch, such articles as a sword, a *tsamba* bowl, a chicken, a pipe, and a pine-torch. From his belt hang flint, steel, and snuff-box, and besides the things mentioned he always carries tobacco and a bag of *tsamba*.

He wears his matted hair done up in a queue bound round the top of his head, and frequently ornamented with a number of coarse silver rings each set with a big turquoise or coral; indeed the Tibetan men are very fond of jewellery and wear several such rings on their fingers. The women however do not make such a display, finger-rings being little worn and ear-rings small. The commonest piece of jewellery is a silver brooch, made in two pieces fitting into each other, which fastens the collar of the jacket.

More curious still, the men always wear threaded on the queue a section of an elephant's tusk, and I found myself wondering why it was that both the Tibetans and Lissus wear characteristic ornaments which are brought from other countries than their own, the Tibetans these pieces of elephant's tusk worn in the hair, and the Lissu women a head-dress, in shape something like a baby's sun-bonnet, covered with cowries, as shown in the photograph. There is little doubt that this last is peculiar to the Lissus, and eminently characteristic of them.

In country districts the Tibetan women plait their hair into numerous thin tails which hang down behind and are collected together into one queue below the waist; but in the cities they wear the ordinary queue, quite half of which is artificial, piled on the top of the head and finished off with two tassels of red or green silk bound with silver wire.

But perhaps it would be more correct to speak of the Tibetans of the Mekong valley, rather than of the province of Kham, for they differ greatly in different parts of this country. At A-tun-tsi is one type, at Pang-tsi-la another quite distinct type, and at Batang yet a third, though it is evident that all these people are nevertheless Tibetans, and resemble each other more closely, for example, than the

Plate X

An Oasis in the Mekong Valley arid region north of Yang-tsa

Tibetan traders at Wei-hsi

Lutzu resemble the Mosos, or either of these tribes the Chinese.

At Yang-tsa a woman asked me for medicine, and after one of the soldiers had told me what was the matter with her, and I had taken her temperature, I gave her certain tabloids and directions with magical effect, for next morning she was very much better and came out to my tent to thank me. This young woman was twenty-three years of age and had three children, the eldest being just seven.

The arid region which we now entered owes its existence to a sudden great elevation of the Mekong-Salween watershed, a very fine triple-peaked snowy range being visible to the west, a little north of Yang-tsa. It is curious that Prince Henri d'Orléans who saw this snowy range in 1895 should call the mountain Doker-la, a mistake which has even been perpetuated by the Catholic priests, who live on the spot. There can be no question that it is a mistake, for Mr Edgar and I independently obtained from the Tibetans the name K'a-gur-pu for this splendid mountain, and I myself crossed the sacred mountain Doker-la, which is not covered with eternal snow, and is correctly placed south of the snow range on Major Davies's map of Yunnan.

Major Davies says that he caught sight of these same snow mountains from the Mekong below A-tun-tsi, and that Captain Ryder calculated their altitude at about 20,000 feet. I saw them from the same point however, and noted that the highest pyramid is not visible from here, so that I think the height will eventually prove to be more like 22,000 feet. This great snow range condenses round itself most of the moisture which crosses the Salween from the southwest, thus acting as a rain-screen to the Mekong valley and to the next range of mountains to the east—the watershed between the Mekong and the Yang-tze. Hence results the arid region of the Mekong valley north of Yang-tsa with a rainfall of only a few inches—probably less than ten— a year, and a greatly curtailed rainfall on the Mekong-Yang-tze divide with a consequent elevation of the snow-line.

The arid region is quite devoid of trees, only scattered rock plants, stunted bushes of *Sophora viciifolia*, and a few other shrubs maintaining the fight against drought, but there was at this season hardly anything in flower, a few *Crassulas* being all I noticed. The rosettes of *Selaginella involvens* grew in their thousands, most of them rolled up into balls like hedgehogs, showing the shimmering under-surface of their leaves, and the fern *Cheilanthes farinosa*, with leaves brilliantly silvered on the under-surface and also capable of curling up, was conspicuous.

Looked at from a distance the arid valley now presented a series of desert mountain slopes and spurs, only relieved by an occasional oasis of green wheat-fields and shady walnut trees, from behind which peeped out here and there big white houses. These oases occur wherever a mountain torrent has built out an alluvial fan capable of being successfully irrigated, since there is practically no water available for crops other than that brought down by the streams, which are few and far apart.

During the halt for lunch on the third day an old Tibetan came out from the village, and besought me to go in and attend to a child who was very sick. I must admit I did not like prescribing for babies of such tender years. But the poor people had always such obvious faith in my powers, and begged so hard for help, bringing me eggs and milk as presents, and thanking me on their knees whenever I did anything for them, that to refuse seemed more cruel than to do the wrong thing.

I found the baby, who was only three years of age and looked less, lying naked on a bundle of filthy rags in one corner of the spacious room, which however, being almost devoid of windows, was very dark. A number of men and women squatted round the fire in the middle, talking and eating. Never shall I forget the awful sight presented by the child as he lay limply on his back, his tiny fists clenched, his mouth half open. Naked did I say? This was hardly true, for as a matter of fact he was almost clothed with flies, which buzzed round in their hundreds as we approached. I counted five inside the child's mouth alone, yet he seemed absolutely indifferent to what would have driven a white child crazy in half an hour. The room, indeed, swarmed with flies, for it was a hot day, and the filthy state in which these people live and keep their houses attracts millions of them—a condition to which the Tibetans seem absolutely indifferent. My patient appeared half starved, so thin and puny that he was almost a skeleton, and on enquiry I learnt that he immediately rejected any food the parents gave him—a fact which I soon verified with a little hot milk. Moreover he was feverish, and not knowing what to do, though the complaint was evidently intestinal, probably dysentery, I gave him a little weak chlorodyne, leaving some at the house with instructions as to feeding him on boiled milk. But I never heard whether the child recovered.

This case set me thinking. One seldom sees small Tibetan children, and I began to realise that the infant mortality amongst these people must be something phenomenal. Only the very hardiest can stand this exposure to all the vile diseases carried by flies, and to the rigors of winter in the mountains, and hardy babies develop into hardy men, who, without flinching, stand pain such as would make the European gasp. Perhaps here too we have one reason for polygamy, so common amongst the Tibetans, for it is ridiculous to assert that the so-called immorality of various Asiatic peoples has no other foundation than immorality for its own sake. Undoubtedly there are several predisposing causes which make for polygamy in Tibet, and this frightful infant mortality may be one of them, since a man is quite uncertain whether a woman can possibly bear him even one child who will survive, and he therefore mates with several women.

Another reason commonly given, not for polygamy but for polyandry, which is also extensively practised in Tibet, is the avoidance by this means of dividing up the wealth of a family. Thus several brothers marry one girl, each living with her for a month or more at a time and signalising the fact that he is in possession by hanging his boots up outside the door!

But undoubtedly the fundamental reason for polyandry is the fact that the Tibetans as a race are only half-way between a nomadic pastoral people and a settled agricultural people. The men are great travellers and leave their wives behind for months at a time, and these good folk solace themselves as best they can with other travellers.

The cynically-minded will add that the argument applies equally well in favour of polygamy. The Lutzu however, who come much into contact with the Tibetans, and all the other tribes of this region, so far as I am aware, are monogamous, which adds the weight of negative evidence in favour of the above theory, since the tribes are notorious stay-at-homes.

We slept at Chia-pieh on the third night, and next day passed through a dismal defile, overhung with rugged cliffs. One of the soldiers informed me that during the Tibetan rising of 1905, four hundred Chinese soldiers who were coming up from Wei-hsi to the relief of A-tun-tsi were ambushed here, and slaughtered to a man, the Tibetans, hidden up in the cliffs, rolling rocks down and hurling them into the river. There must be some truth in the story, but the numbers are probably exaggerated. Talking of fighting, curious tall watch-towers of mud, square in plan and extensively loopholed, frequently occur on prominent eminences near the villages, and there are more of them on the Yang-tze. I believe they date back only some fifty years, having been built by the Chinese when they subdued the Mantze tribes of Yunnan and Ssu-chuan in the neighbourhood of the main roads.

Leaving the Mekong behind, we ascended the narrow valley towards A-tun-tsi. The stream higher up was lined with tamarisk bushes, and just before we reached the city we had a glorious view of K'a-gur-pu, its triple peak blocking up the entire mouth of the valley, and rising apparently to a prodigious height. But to launch out into superlatives when one has not the most prosaic instrument for the measurement of angles is unwise.

The first person I met in A-tun-tsi was M. Perronne, a French gentleman who was there for the season buying musk from the Tibetans. He had been in A-tun-tsi during the summer months for the last six years, knew everybody of any importance in the village, and was always extremely hospitable, so that when in A-tun-tsi I enjoyed many hours in his company, while his knowledge of the language and people was always at my disposal in any difficulty. I dined with him that evening, and learnt a good deal about the village which was to be my base camp for the next six

Plate XI

A-tun-tsi from the South; ruins of the monastery in the foreground

months and in return I gave him what news I could from T'eng-yueh and Europe.

For the next two weeks in splendid sunny weather I explored the surrounding mountains, but in order to make such description of botanical work intelligible, some preliminary idea of the country is necessary.

The village, with a population of about 250 families, is situated at the head of a narrow valley on the lower slopes of the Mekong-Yang-tze divide, the main part of which forms a high rocky range of mountains to the east, its altitude according to Major Davies being 11,500 feet, a higher figure than that given by Prince Henri. The old monastery is in ruins, having been destroyed by the Chinese troops in 1905, but it has since been rebuilt on a hill immediately overlooking the village, which consequently, as far as one can see, is again at the mercy of the lamas.

Considering its extraordinary situation, pinched in the head of this narrow valley, one is not surprised to learn that in the fighting between Chinese and Tibetans, this little village was taken and retaken no less than five times. Of all the horrors then perpetrated we know nothing, though it is freely admitted that the most awful tortures were inflicted by each side in turn when their opportunity came. Five hundred Chinese and Tibetans were killed here alone.

Here is a story from Chun-tien, a city with a big lamasery, eight days' journey to the south-east of A-tun-tsi.

The Chinese official at Lichiang-fu sent two messengers to the head lama, warning the priests not to join their fellow-countrymen in the rebellion, or take up arms against the Imperial Government. The lama's reply was characteristic of the ferocity of the Tibetans when roused, one man being taken and skinned alive, the other sent back to take news of his fate to the Lichiang official. And the monastery still stands!

Since the rebellion, the garrison of A-tun-tsi has been increased to a hundred men, numerous Chinese merchants have established themselves, and the Tibetans, now poverty-stricken, have been deprived of all power. Most of the old lamas were either killed during the fighting or executed afterwards, very few escaping into the mountain fastnesses of Tsa-rüng.

At Batang, His Excellency Chao-Er-Feng, Warden of the Marches, who was recently executed by his own soldiers, was accused of committing numerous atrocities, beheading men, women, and children indiscriminately when the day of retribution came.   This may or may not be true, but one thing is certain, namely that, under the circumstances in which he was placed, Chao acted on the whole humanely, according to his lights, on behalf of Chinese and Tibetans alike.   He beat to death more Chinese soldiers for double-dealing with the Tibetans than he did Tibetans for murdering the Chinese.

Personally my sympathies are all with the Tibetans, whom I like, but one must never forget, in judging the Chinese methods of repression and retaliation, that they were dealing with a barbarous and scattered people covering an immense tract of inaccessible mountains ; that these people are hardy, resourceful, and unimpressionable ; and that they had wantonly put to the sword the Chinese Amban at Lhasa, one holding the rank of Viceroy—a position in the Chinese Empire second only to that of the Emperor himself—together with over a hundred of his followers.

Between A-tun-tsi and the Mekong to the west is a high mountain rising in steep screes and precipices devoid of vegetation to a rocky ridge about 4000 feet above the village ; and when the summer rains break, great quantities of gravel and mud are washed down the valley, inundating the houses at the lower end of the village.   Landslips of this nature occur periodically, every year mountains of mud descending and doing no little damage.   In August 1911 it rained furiously for a week and a tremendous quantity of the mountain came away, forming a river of liquid mud which damaged several houses and made a complete wreck of one.

The deep valleys which seamed the high range to the east were clothed with vegetation up to 15,000 feet, above which altitude south-facing slopes, exposed to both sun and wind, presented bare screes with a scattered and highly-specialised flora, while north-facing slopes were covered with turf, forming what may be called alpine grass-land. In the narrow valley itself below 12,000 feet there was no real forest, the slopes rising immediately above the village being clothed with scrub oak and numerous rhododendrons, many of them now in bloom, with a variety of shrubs and small trees, such as the mock orange, barberry, *Hippophaë rhamnoides*, cotoneaster, and *Deutzia discolor*. *Indigofera*, *Caragana*, *Desmodium*, and other shrubby Leguminosae were common, with *Spiraea canescens* and innumerable roses, and the slopes were covered later with masses of white and yellow clematis, chiefly *C. montana* and *C. nutans*.

The forests consisted mainly of firs, but in places there were poplars, maples, and birches, and here too were a few climbers of the genera *Schizandra*, *Vitis*, and *Actinidia*. The beautiful pink-flowered *Podophyllum Emodi* grew in the shady nooks of the shrub belt, along with two delightful Cypripediums—*C. Tibeticum* and *C. luteum*—and higher up, at 13,000 feet, the white-flowered *Souliea vaginata* was just blooming, its flowers opening almost before the leaves appeared, as is the case in so many of these shade plants. In drier places was a yellow-flowered shrubby paeony (*P. Delavayi*) and a dwarf blue iris (*I. kumaonensis*). Several species of Primula were already in flower, mostly shade and moisture-loving species, such as *P. sonchifolia*, the recently discovered *P. lichiangensis*, and *P. septemloba*.

At 15,000 feet there was still a good deal of snow, at least on north-facing slopes, and consequently not much was in flower as yet.

At this season the Tibetans, who still form more than half the population of A-tun-tsi, exclusive of soldiers—for nearly every Chinese soldier has a Tibetan wife—were preparing with child-like glee for their annual festival, which fell on the first three days of June.

The Tibetans always strike me as being so much more jolly and irresponsible than the Chinese, who are just as sedate and gloomy when enjoying a holiday as at any other time. The latter never seem able for a single moment to shake off the idea that life is a serious fever which has to be borne, and consequently they never take risks. Every question of policy is carefully debated from all possible points of view before any particular line of action is decided upon; everybody asks everybody else's advice before committing himself, and nobody ever does anything unless he has a majority behind him. Genius branching out on its own responsibility, if it ever occurs, is severely frowned down ; there is no room for sentiment. This communistic principle in village life is so frankly at variance with the one-man autocracy of higher administration that it would be surprising in so practical a people, did one not realise that inconsistency is characteristic of the nation. Each rank is consistent with itself and that is all that can be said. Nevertheless this practice enables the villagers to suggest simple remedies for all evils, and inculcates in them a great capacity for managing their own affairs without falling back upon authority in the shape of the nearest mandarin.

Here is an example from A-tun-tsi.

One of several treasurers for the sum of Taels 1400 (about £180) subscribed by the Mohammedans for the purpose of building a mosque showed a defalcation of Taels 300 when the time came for the money to be produced, and the brotherhood of local merchants, instead of appealing to the mandarins, of whom there were two on the spot, thrashed the matter out among themselves. All day they sat in conclave, smoking and talking together, and at the end of the day, having examined the defaulter and several witnesses, they unanimously decided that the money must be paid back by monthly instalments as soon as possible, the merchant's furniture and effects being sufficient surety for him. Thus his peers were witness, judge, and jury.

The Tibetan festival itself seemed more in accord with the usages of *Nat* propitiation than with Lamaism, except that it was eminently cheerful, and the people, led by their priests, went to the summits of the three nearest hills to east, north, and west in turn, in order to burn incense and pray ; after which they ate cakes. The first day however was devoted entirely to the amusement of the children, for Tibetan mothers, as I frequently observed, are warm-hearted creatures with a great affection for their offspring.

Dressed in their best frocks, and wearing all the family jewels brought out for the occasion, they went up into the woods in the afternoon, picked bunches of flowers just as

Plate XII

Tibetan women of A-tun-tsi

Chinese and Tibetan boys in holiday attire at A-tun-tsi

English children love to do, romped, made swings and swung each other, and finally sat down to eat cakes, which they had been busily making for a week past.

Just as the young of different animals more nearly resemble each other than do the adults, so too are children very much alike in their games the world over ; picnicing is not confined to Hampstead Heath, nor picking flowers to botanists.

In the evening they all trooped back to the village to dance in the mule square, and skip. Three or four little girls would link arms and facing another similar line of girls advance and retreat by turns, two steps and a kick, singing, in spite of their harsh voices, a not unmusical chorus; the other side would then reply, and so it went on, turn and turn about.

It was a most delightful parody of that pretty little Christmas game " Here we go gathering nuts and may," and I enjoyed watching it though I did not understand the words, which were probably less ingenuous than in the ditty quoted. But the girls themselves, in their long frocks of dark blue cloth buttoned up one side and trimmed with a narrow border of white, long-sleeved jacket to match, scarlet cloth boots, and tasselled queues, looked charming. All wore several silver bangles, besides ear-rings and large brooches, practically all the jewellery they could find room for. The boys played together, but were less resourceful than the girls and, as in other parts of the world, never seemed to know quite what to do with themselves. They wore smart white coats for the great occasion, but favoured Chinese dress, and probably they are made to do so in the schools in A-tun-tsi, for one never sees a small boy belonging to the village in Tibetan dress. One boys' game is however worth mentioning. They called it 'eggs' and it is played as follows. One boy is in the middle—a fundamental necessity in nearly all children's romps—and sprawled on all fours over several pebbles, representing a bird sitting on a nest of eggs, which the others, who danced round, were trying to despoil. When a favourable opportunity offered itself, one of the pillagers would dart at the eggs, and if he secured one without being kicked or hit, he was deemed to have been successful ; otherwise the mother bird who pivoted round with much agility, kicking out right and left, in front and behind, put him temporarily out of action—metaphorically speaking.

My landlady, whose young sister was one of the revellers, gave me a plateful of dainties—little round cakes made of pastry with a great deal of pig's fat in it, filled with chopped up walnuts and spices—and in the evening, having lighted up the family altar with a row of tiny brass butter lamps, she placed other little cups full of grain in front of the gods—again recalling *Nat* propitiation; for these offerings to the deities are no part of true Buddhism I believe, and the debased Lamaism of Tibet really seems to have been grafted on to an original *Nat* worship.

On the last afternoon I went up to the lamasery on the hill to watch the service for half an hour, but it is not a very wonderful place and in its internal decoration boasts only the relics of its former glory. The one living Buddha or reincarnation sat cross-legged on a dais at the far end of the temple, and the seventy or eighty priests present, many of them only small boys, sat on the floor in rows, mumbling the responses. From time to time the band struck in with a wild crash, the wail of the conch-shells, the jingle of cymbals, and the blare of the long trumpets sounding very strange in the big dimly-lit hall. Many people were, like myself, standing round looking on, and the younger priests at least were more intent on watching the crowd than on their devotions.

Presently, on a signal from the usher—an oldish lama with a wrinkled face continuously wreathed in smiles, who stood just behind the Buddha's throne—the people began to file round the temple, each as he passed in front of the Buddha placing a small strip of white silk on a pile already gathered, and receiving in exchange a blessing which consisted of nothing more material than a tap on the head from the left hand of the Buddha, palm upwards; and they had to pass before him in an extremely uncomfortable doubled-up position to get that. The pilgrims for the most part retained their gravity till they were half-way down the temple again, when they began to giggle in a most unseemly manner, evidently in contemplation of the trick they were about to play on the old man; for at least half

of them joined the queue again and went round a second
time, thus receiving two blessings for one piece of silk,
and like Esau needlessly squandering the poor old man's
patrimony. A number of children who went round with
the procession had nothing at all to offer, though they too
received a perfunctory blessing and moved on, highly
pleased with themselves.

On June 2 I called on the official and asked for permission to return to Tsu-kou for three weeks, as there were no
flowers yet on the high mountains round A-tun-tsi. I had
worked the lower slopes pretty thoroughly during my
fortnight's stay, and had not only found several nice things,
but had observed that in another month I ought to make
some rich finds. In the meantime I was anxious to cross
the Mekong-Salween watershed and prosecute my search
westwards.

Chao, the chief official, offered no objection to my
returning to Tsu-kou and promised to send round the
necessary animals on the following morning. Sung was
therefore deputed to remain behind and look after my
baggage, while Kin and I, with food for a week, tents,
medicine chest, camera, and a few other necessities set out
on the morning of June 3.

CHAPTER VI

A JOURNEY TO THE SALWEEN

THE return journey to Tsu-kou occupied just over three days and was uneventful. The Mekong gorge was becoming more and more parched now, for the warm summer winds were already blowing with daily increasing fury as the deep trench became heated up, and the cold air swept down from the overhanging mountains to take the place of that dancing up from the scorched rocks.

The nights however were delightful. My tent was always pitched near some thundering torrent, which was music to me, and lying in bed I would watch the brilliant arc of the new moon set over the mountains. I was travelling with an *ula* passport obtained from the yamen, my men commandeering from each village in turn the requisite number of animals and porters, who were bound to take me as far as the next village, the rate of pay being fixed by the yamen. This system of relay transport obtains on the main road across Tibet and in the Tibetan Marches for the convenience of Chinese officials travelling on business—the Tibetans, nominally in return for grants of land, agreeing to keep in readiness a certain number of transport animals at fixed posts. As far as cheapness and certainty of transport are concerned, the system has its advantages, but from other points of view it is often very unsatisfactory, an hour or two being sometimes wasted changing animals when others are not immediately available. When this happens several times a day, it becomes exasperating.

In the Mekong and Salween valleys the loads were generally carried by women and boys, and in all my

Plate XIII

My Caravan coming through the Mekong gorges on the way
to the Salween

journeys, when travelling with *ula*, I found that the people
were more ready to carry the loads themselves than to find
the necessary animals; probably because the country is so
poor and offers so little grazing land near by, that transport
animals are scarce. Cattle, it must be remembered, are,
with the sheep, taken up to the high alps for the summer.
Mules are little used in S.E. Tibet; ponies are few even
in large villages like A-tun-tsi; and donkeys, chiefly used
by the big lama caravans making pilgrimages to Lhasa, are
difficult to procure.

At Yang-tsa I saw the woman whom I had doctored on
the way up; she was now quite recovered from her fever,
and greeted me in the most friendly fashion, inviting me to
lunch in the large dark kitchen, where I saw a new method
of warming tea. Several flat stones, each with a hole bored
through the centre, were placed in the fire till they were
red hot, when they were hooked out and dropped one by
one into the wooden tea-churn, in which we could hear the
tea bubbling merrily.

One night I camped to leeward of a magnificent
Asclepias tree in full bloom, which perfumed the night
air. The cicadas were chirping all round, the Mekong
could be heard thudding over the rocks below the village,
and so brilliant was the moonlight that I was able to write
my diary in the open air. And sitting there, I heard above
the voice of nature yet another sound, the sound of drums
being wildly beaten. Presently there came into view a
procession of villagers headed by priests, carrying fragrant
torches of pine-wood, sticks of smouldering incense, drums,
gongs, and red sign-boards on which various Chinese
characters were depicted. Outside the tiny village temple
an altar had been erected, and a fire crackled lustily at
the entrance, sending a dense column of smoke up into
the air. On the altar were trivial offerings of grain to
the gods, and no doubt many paper prayers were burned
and wafted to them on the breeze. Then the procession
marched round and round the altar, and passed into the
fields, waving the torches madly, while the noise grew
louder. They were praying for rain. The whole ceremony
was a strange mixture of Chinese and tribal customs, for
there was more than a hint of *Nat* propitiation in it.

The river had risen considerably during the past three weeks and the water was no longer olive green but chocolate red.   Still there was no hint of rain till we entered the rainy belt just above Tsu-kou.   A ribbon of blue sky followed the windings of the river ; the wind gradually rose each morning, raged with increasing violence throughout the middle of the day, and died quietly away towards evening ; the bare rocky slopes glared in the sunshine and danced in the quivering air.

One morning I picked up one of the large, brilliantly coloured bugs (*Hemiptera*) which swarmed on the flowering shrubs now in full bloom, but he emitted such a powerful odour that I quickly put him down again.   At the same moment a fly got in my eye, and I thoughtlessly put up my finger and began to rub it.   Great heavens !   I began to think I would never see again.   It was as though a drop of vitriol had been squirted into my eye, which for some minutes ran with water and gave me great pain, and though there could not have been more than the merest trace of this excreted liquid on my finger, it was evidently a most powerful acid, a drop of which would doubtless blind a man.   After a time the pain passed off, but for the rest of the day my eye was puffy and bloodshot.

Early on the morning of June 5th we reached the rope-bridge at Tsu-kou, and crossed to my old camping ground in the gulley ; and I lost no time in seeking Gan-ton and telling him that we must start without delay.   He therefore promised to find porters and get off on the following day as quickly and quietly as possible.

That night I saw fire-flies and listened to the bull-frogs groaning down by the torrent, but next morning it was raining again, and the limestone precipices which crowned the first range were buried in mist.   I spent the morning climbing up the gulley to the foot of the gorge, finding several beautiful orchids in flower on the limestone rocks. I saw, too, several common pheasants, and heard snow-pheasants calling in the bushes.   Fine butterflies, including the English swallow-tail, *Papilio machaon*, or a closely allied species, inhabit the valley, and they became quite a nuisance in my tent, settling on the milk and butter tins in clouds. On this excursion I also saw a small black snake with an

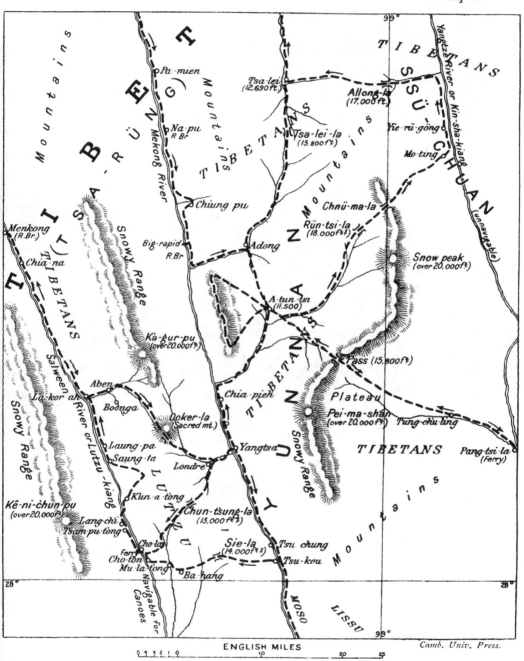

Map IV

Sketch map to illustrate journeys (i) round **A-tun-tsi**, (ii) to the Salween, and (iii) to the Yang-tze.

[*Chapters V to IX: XI to XV.*]

[*R.Br. = Rope bridge*]

expanding hood, like a cobra's, and a big water-snake, the
only two I recollect seeing in the year, though lizards were
common enough on the dry rocks of the arid region.

At mid-day Gan-ton came along with his six Tibetan
porters, and after lunch we started up the mountain in the
rain. Fate was on our side, for the small military official
posted at Tsu-chung was away, and there was no one to
hinder us.

Our party consisted of Gan-ton, who was guide and
interpreter, Kin, six porters, my watch dog Ah-poh, and
myself. Ah-poh was a Tibetan mastiff, and he became
much attached to me, though the most I ever taught him
was to come when I called. Ascending the barrier range
through pine-woods, we reached the last village, situated
among steep stony slopes, from which a thin crop of barley
is extracted, and at the topmost hut stopped for the night.
We were not far below the pass over the spur, and perhaps
2000 feet above the Mekong.

Throughout the night it rained steadily. I had pitched
my tent on the flat roof of one of the huts, but unfortunately it was not flat at all, but sloped in towards the
centre, and when I awoke next morning the inside of the
tent was a lake, into which I inadvertently stepped from
my bed.

It continued to rain all the morning, and the porters
declared that it was quite impossible to proceed under
these circumstances, for the path was steep and slippery;
so I endured the delay with what patience I could, and
spent the time draining my tent. After lunch the rain
held off for a bit and we started again, crossing the spur
(9000 feet) and descending into the dense rhododendron
forest. Towards the summit, clumps of *Cypripedium
luteum* were prominent, with a very large white-flowered
Viola, a *Rodgersia*, an *Osmunda*, and other shade plants.
A steep and slippery path led down from the pass towards
the bed of the torrent, and then began the long pull up
the valley, tripping over roots and creepers, clambering
over logs, and slipping this way and that in the mud.

I have already compared the dry, pine-clad slopes of
the barrier range fronting the Mekong valley with the
richer vegetation of the gullies, and the deep valley behind

the barrier ridge in which we now found ourselves presented forests and undergrowth richer than anything we
had yet come across.

Amongst the dense thickets of rhododendron, many of
them big trees thirty feet high, but none now in flower, a
magnificent spruce was conspicuous, besides maples, birches,
and alders.   Between the forest trees and the herbaceous
undergrowth was a second tier of shrub vegetation, Caprifoliaceae (*Lonicera, Viburnum, Diervilla*), Saxifragaceae
(*Ribes, Deutzia*), Rosaceae (*Rosa, Rubus, Prunus, Spiraea*),
with *Clematis montana* and clumps of bamboo.

The undergrowth itself varied considerably according
to the nature of the forest, consisting in one place mainly
of ferns, in another of a yellow-flowered Corydalis growing
very rankly, but just here it was far more varied. A
pretty pink Oxalis was conspicuous, with several species
of Arisaema; here and there was a Listera, Monkshoods
occurred in quantity, and numerous plants of *Lilium
giganteum* were just showing their leaves.   Indeed, the
Liliaceae were as richly represented as any order, and
included some very sweet-scented Oligobotryas both pink
and white-flowered, Paris, Convallarias, and others. There
were ferns too, as stated, Aspidiums and Aspleniums chiefly,
but these were confined to the damper parts of the forest.

The Aroids (species of Arisaema) were the most peculiar
of all, for many of them had not only the spathe drawn
out into a flagellum several inches in length, and covered
by a lid very similar to that of Nepenthes, but the spadix
itself was similarly attenuated.   Instead of being straight,
however, the spadix was bent sharply down on itself, by
which means it was brought out of the investing spathe,
and then turned upwards again in its original direction.
The object of this external spadix, which is coloured, is
undoubtedly to attract and guide insects to the flower, the
double bend being merely a mechanical device to expose
the spadix without the necessity of raising the lid. As to
the latter, which is simply the triangular tip of the spathe
bent over and drawn out into a long flagellum, this is
perhaps useful on account of the persistent rains in the
forest where these Arums grow; but the use of the
flagellum itself, which is the most remarkable feature of

Plate XIV

The Mekong-Salween Divide, seen from below the Sie-la

the plant, is not so easy to determine, though by its mere weight it holds the elastic flap well down over the aperture and effectually closes it from above, leaving only a narrow slit round the edge. This slit, however, is not very accessible to insects, owing to the rim of the spathe being turned down and outwards, and it seems to me that an insect's only line of entry is down the spadix.

If indeed the flagellum is affected by the weather in such a manner as to raise or lower the flap according to the humidity of the atmosphere, we should have its use clearly defined : but though I am inclined to ascribe some such function to it, I never saw it operate in this way, for it was always raining when I was in these forests.

As we climbed higher, mixed forest gave place first to birch and finally to open alder forest, with a dense undergrowth of Corydalis : rhododendrons, spruces, and other trees had all disappeared from the valley bottom, though higher up they still clothed the mountain slopes.

Here we found a small wood-cutter's hut, and into this we crowded just as the rain began again, making the night very chilly, for it poured steadily throughout it, and dripped through the shingle roof, so that with daylight I was glad to crawl from beneath wet blankets into wet clothes. The prospects for the day looked utterly miserable, but at seven o'clock it began to clear up, and when we started an hour later, the rain had ceased. Passing through the drenched alder forest which grew rapidly thinner, we presently emerged into the open, where alpine meadow, with tall grasses and flowers, alternating with clumps of bamboo brake and fir trees, clothed the valley. Now we waded the cold stream (at least the men did, but I was carried across), recrossed it higher up by means of a tree-trunk which afforded precarious foothold, climbed steeply up through bamboo thickets, and, crossing once more over a snow bridge beneath which the torrent had cut its way, reached the head of the valley and sat down to lunch. Before us rose the snow-clad rock wall of the Mekong-Salween watershed.

The alpine flowers spread out at our feet were magnificent—the glorious sulphur-yellow *Meconopsis integrifolia*, *Primula sikkimensis*, large Thalictrums, Caltha, Anemone,

Primula Watsoni, and higher up the dwarf blue *P. bella* with Soldanella, crimson azaleas, purple columbines and many more, forming sheets of colour.   Every rock and boulder supported a small garden of saxifrages and tufted alpines, every marsh displayed masses of some rare flower such as *Primula Souliei*, with sedges, gentians, and sphagnum.

Here and there streams coming down from the mountain in front had been ponded back by detrital cones washed across their path, and had consequently formed bogs from which I think a good peat might be dug.   Thin cascades came tumbling over the precipices in front of us, and in the smooth rounded rocks which here occupied the valley floor and the piles of scree or moraine material through which the torrent had cut its way, there were indications of previous glaciation.

For nearly three hours we struggled up through the soft snow, frequently sinking in to our knees, the porters even to their waists, requiring to be pulled out by their companions.   Finally we reached the summit, and lighting a fire, I boiled some water, which I found boiled at 85·5°C. giving 14,500 feet as the approximate height of the pass. This however is probably too much, and 14,000 feet may be nearer the mark.

At the top of the pass the snow was banked high, and ended abruptly, the southern face, which was extremely precipitous, being clear for the first 1000 feet or so, though a single large drift lay down below.   A cold wind was blowing up either side of the pass, and here we have an explanation of the fact that while the slopes of the col itself were bare of trees, to the right and left on the neighbouring peaks spruces extended upwards to a height above that of the actual pass ; but there they were not exposed to this concentration of wind along the valley bottom, which raked the col from both sides.

The Sie-la as the pass is called, has been crossed several times by French officers and travellers, and is frequently crossed in all weathers by the intrepid Catholic priests of the Mekong and Salween valleys.   It is said to be clear of snow for two or three months in the year, though when I crossed it in November it was already more deeply under snow than in June.   Were it clear of

Plate XV

Porters crossing the Sie-la, 14,000 feet

snow, there would be no great difficulty in taking mules across, though it would be a slow business, but with the snow as deep and soft as we now found it, the undertaking would be hazardous, and I was glad we had not attempted it.

We spent half an hour resting on the summit. There was no view, though to the east the snow-clad Mekong-Yang-tze divide showed up, while to the west we looked only into a deep valley. Behind us lay China, in front Tibet.

The pass faced E.N.E. and W.S.W., presenting considerable differences of vegetation on the two slopes, due to the earlier melting of the snow on the south side, which was consequently much drier, besides being steeper. It was covered with a short alpine turf dotted with hundreds of flowers of the small mauve *Primula pulchella*. Later in the summer, other flowers appear, including a *Meconopsis*, which I found in seed five months later.

On the snowy slope, however, were shrubs, chiefly azaleas, junipers, and willows which protruded through the snow here and there, brown tussocks of a fern with no new fronds showing as yet, and the silvery leaves of a potentilla. Evidently this slope is covered with a rich assortment of plants from July to September, but when I saw it again the snow was deeper than ever and not a twig showed through.

The rock appeared to be a grey banded gneiss with a good deal of quartz, much crumpled and tilted, so as to present sharp edges above and steep precipices below.

Descending from the pass, we found ourselves in a larger valley running more or less north and south, and passing through alpine meadow once more, we reached a second hut on the edge of the forest towards dusk; the rain which had held off all day setting in again. All round us was high alpine meadow with scattered willow, alder, and birch trees draped with lichen, which flapped dismally in the wind. A tall crucifer (*Cardamine macrophylla*) was conspicuous here, and Gan-ton informed me that the Lutzu and Tibetans boil and eat the leaves and stems. He also pointed out some brown fritillaries and a hyacinth, the bulbs of which are eaten by the natives, an Umbellifer of which the leaves are eaten, and several edible toadstools, besides aconite used to poison their arrows.

In the evening two Lutzus, who were crossing the mountains to Tsu-kou, arrived at the hut and joined my men round the fire. They had with them several voles (*Microtus Wardi*) which they had trapped higher up, and these they proceeded to skin and fry for supper. I ate one myself, and must admit it was not at all bad. Gan-ton told me that the Lutzus make special bamboo traps for these voles, of which they catch and eat a great many. I saw one of the traps later, an ingenious contrivance, on the principle of the pheasant-trap already described, the vole crawling out of his hole and putting his head into a noose, which is immediately drawn tight by the springing up of a bamboo.

Next morning we climbed straight up out of the valley over the western ridge, passing through dense groves of bamboo brake, and higher up through spruce and rhododendron forest. From the pass (water boiled at 87°C. giving an approximate height of 13,000 feet) we had a clear view westwards right across the deep Salween valley to the Salween-Irrawaddy watershed beyond, but unfortunately it was extremely cloudy. What we did see, however, was this.

A range of snow-clad mountains stretching from south to north rose above the cloud-banks, and high above all one peak stood out conspicuously, its snow-fields and glaciers seen distinctly through the moisture-laden atmosphere. This is the mountain known to the Tibetans as Kê-ni-ch'un-pu, which rises right above the village of T'sam-p'u-t'ong on the Salween. Beyond this range we come to the Nmai-kha or eastern branch of the Irrawaddy, where reside the Ch'utzu, a little known tribe who, I was told, tattoo their faces, though I do not remember that Prince Henri remarks on this in his book, *From Tonking to India*, and no one else, unless it be Captain Bailey, has been amongst them.

Judging by the views I obtained of this range in November from another pass, it does not appear to be covered with eternal snow for any great distance either north or south of the main peak, though it evidently receives the full force of the south-west monsoon blowing

Plate XVI

At the limit of trees, Mekong-Salween divide: Abies forest thinning out at 13,000 feet

over the plains of Assam, thus acting as the first great rain-screen to the deep valleys we are considering. Consequently there are probably no higher ranges to the west in this latitude, and the snow-line on the Salween-Irrawaddy divide is likely to be even lower than on the Mekong-Salween divide.

Captain Bailey, who in the same month crossed to India further north, mentions seeing from above the Irrawaddy a snowy range to the east of him—probably Ta-miu between the Mekong and Salween, to which I shall refer later. He then crossed a pass between the Salween and the Irrawaddy 15,676 feet above sea-level, where there was a good deal of soft snow on June 17th. The pass, however, would probably be clear later, and judging from his account, I gather that this divide is not so lofty above Men-kong as where I saw it further south.

What, however, struck me more than this wild snow-clad range peeping up above horizontal strata of silvery cloud, was the appearance of quite another range running more or less at right angles to the above—that is east and west—but as I was looking at it end on, it was difficult, on account of perspective, to be certain of the direction of this range, while the fact that the far west was buried in cloud made it still more deceptive. Afterwards I learnt that the late Mr Lytton believed he had seen such a range from the Salween valley further south, so that there is some authority for my statement, but on the other hand Captain Bailey says nothing about a transverse range, and he, if anyone, should be in a position to know.

Does this cross-range, terminating the Tibetan plateau in the form of a high bluff, really exist? If so, it seems probable that the eastern branch of the Irrawaddy at least rises from its southern slope. No better natural boundary between Assam and Tibet could be devised.

Near the pass, the grassy hill side was honeycombed with vole burrows and I noticed one or two traps, but I did not see any voles scampering about as the pica-hares do by day on the Tibetan plateau.

Descending through dense fir forest, the path lined with a fine mauve Primula (*P. nivalis*), we emerged on to open grass-covered hills where yak and cattle were grazing.

Patches of scrub showed up below, and the gullies were choked with dense forest. From time to time the sun broke through, and gleamed on a high yellow cliff far below us, which looked strangely small against the dark range behind. This limestone bluff marked the position of T'sam-p'u-t'ong.

The valley immediately below us was not that in which the Salween flowed, for there was still another barrier ridge to cross. I do not quite understand the conditions necessary for the formation of these parallel barrier ridges— spurs from the main divide—but they are probably due in part to the arrangement of the rocks, though, as previously suggested, the distribution of rainfall may also have something to do with it. I never saw a mountain stream flowing parallel to the main river in the arid regions, though in the rainy regions of both the Mekong and Salween it is the normal thing.

In the evening we reached Bahang, or, as the Chinese call it, Pei-han-lo, a small Lutzu village perched on the side of the mountain more than a thousand feet above the valley. There is a French Catholic priest stationed here, but he was away; also a posse of Chinese soldiers not very far from the village.

No sooner was my tent up than it began to rain, and continued all night and all next morning. Accordingly it was impossible to start in the morning, and when we did get off in the afternoon the clouds were very low and it was still drizzling, so that we only descended to the valley, and continuing up it for a mile or two, took shelter in a Lutzu hut in the village of Mu-la-t'ong just as the rain began again. Gan-ton said it rained every day in the Salween valley, and so far I had had no occasion to disbelieve him.

No sooner were we in our new quarters than a woman came to me for 'foreign medicine,' and it was now that I congratulated myself on having brought my tabloid medicine chest with me, for here was an opportunity to get on a friendly footing with these shy natives, though it was unfortunate that the patient was another baby. I always found babies the most unsatisfactory patients, for they could not themselves say what was the matter with them and their mothers didn't know, which left the amateur very much in the dark. However with the help of Gan-ton, who resurrected such symptoms as he could from the mother's mind, and a thermometer, we diagnosed the case and prescribed simple but innocuous remedies, leaving the rest to faith—the parent's faith, that is, which was infinite. I found too, after a little more experience of sick babies in this region, that they rang the changes on two or three mild maladies which I eventually became quite expert in coping with.

We were a lively crowd in the one room of the hut. I had a corner to myself where my bed was set up under a small square window cut out of the log wall; there was no shutter to it, but in spite of the incessant rain it was not very cold. Round two fires in the middle of the room were gathered my eight men, the woman of the house, her three children, and four more men. What may be called the floating population, who looked in occasionally and sauntered out again at will, included several of the villagers, a calf, a cat, two or three pariah dogs, and a flock of hens, while the odour and grunts of sundry pigs rose up between the loose floor-boards.

The Lutzu tribe, amongst whom we now found ourselves, are interesting for the reason that they seem to indicate an irruption of tribes from the west. That they have come *down* the Salween valley from Tibet, representing one of the links in a chain of emigration in that direction, I do not believe, and so far as language is any test, the Lutzu tongue seems to bear no more resemblance to Tibetan than could be accounted for by the fact that the Lutzu are a small tribe enclosed by Tibetans who, being great travellers and traders, have long been in and out amongst them. The English language has been influenced in much the same way by Norman, but is not related to it.

The Lutzu on the other hand, are not traders at all, being in the enviable position of having everything they require, hemp for their clothes which are woven by the women during the winter, tobacco, maize, wheat, buckwheat, apples, oranges, and so on. Bamboos and gourds supply them with vessels, and with the cross-bow they shoot game. Though not a drunken people, they certainly drink large quantities of liquor made by fermenting maize;

but this beverage, which is of the consistency of pea-soup and is taken warm, is probably more nourishing than inebriating. In the winter men and women sit round the fire for hours at a stretch, chatting, smoking, and drinking. It is meat and drink and medicine to them, and by no means unpalatable.

The men wear their pig-tails down, not bound on the top of the head as do the Tibetans, and their dress, though simple, is not unpicturesque—short breeches (probably copied from the Chinese) and shirt of white hemp cloth, trimmed round the collar and sleeves with light blue, and strips of cloth wound loosely round the calf, like puttees. The women wear a single long-sleeved garment usually of dark blue cotton cloth, reaching below the knees and tied round the waist, and frequently a hempen cloak, extending across the chest from the right shoulder to the left arm-pit, is added. A hempen bag, decorated with seeds but of plain workmanship, is slung over the shoulder, and it may be remarked that similar bags are carried by most of the tribes west of the Mekong, but not by the Tibetans or Mosos. They usually bind the pig-tail round the head, after the style of the Tibetan women, but there is little jewellery worn. Some of the girls before child-birth are extraordinarily handsome. Their complexion is decidedly lighter than that of the Tibetans, but not so sallow as that of the Chinese, the features are regular, the nose well bridged, the eyes large and round, the high cheek-bones scarcely prominent.

The religion of this people is a modified form of Lamaism, but I believe this has been clumsily grafted on to a much older cult, probably *Nat* propitiation, for in common with the Lissu and other tribes they hang up special corn cobs in their houses, which are, I think, in the nature of propitiatory offerings to the Penates, a practice which is not observed by the Tibetans at all. How far their Buddhism differs from the degraded form current in Tibet, I cannot say, for the only rite I ever saw anyone perform was when the young lady of the house took up a jug of water, and made the sign of the cross over the household fire by throwing water across it to north, south, east, and west. She did it in a very business-like way, just as she might have fed the chickens, the first shower hitting the wall behind, and the last one drenching me.

I find it difficult to escape the conviction that the Lutzu, now essentially an agricultural people, represent a jungle tribe in a comparatively advanced stage of civilisation. Their short stature, their method of carrying loads by means of a strap passing round the forehead, their use of the cross-bow, pre-eminently a jungle weapon for jungle warfare, owing to its short range and diabolical effectiveness, their gourds and bamboo tubes, and their rope bridges, all suggest this.

It is a well-known fact that since the great mountain ranges and deserts of Asia stretch east and west, emigrations on a large scale have, for the most part, taken place in this direction also, it being easier to skirt such obstacles than to cross them. But a narrow gateway lies open southwards through the rampart of mountains which rims Central Asia in the region of the parallel rivers, and, while marvelling at the one corner of high Asia where great emigrations in a north and south direction have probably taken place, one should not lose sight of the fact that emigration may here also have taken place in an east and west direction.

All night long and all the following morning it poured with rain, nor had it entirely ceased when we started after lunch. The morning was spent in doctoring the sick, for my fame as a healer had gone abroad, and at one time we had six mothers with unweaned babies at their breasts asking for medicine, till, with my own men and curious spectators to a total number of thirty, the hut looked like the out-patient room of a London hospital. We had a splendid variety of ills, from a bruised shoulder—the result of a scuffle—to a sick headache, and we did something for all of them. Their gratitude was unbounded, many of them bringing me presents of milk and eggs.

During the day, a big caravan composed of twenty-five animals and about forty porters passed through the village, carrying silver and supplies of rice, sugar, tea, and so on to the troops at Bahang. They had been six days on the road from A-tun-tsi, crossing the mountains by the

Chun-tsung-la, a pass further north, only 13,000 feet high, the lowest and easiest pass in this region, which we were afterwards destined to cross in November on our second journey to the Salween.

The garrison at Bahang was established for the purpose of holding the Lissus in check from the north, and the soldiers make periodic excursions down the Salween valley, there to fight with the tribesmen. These latter are thoroughly accustomed to jungle warfare, at which they are adepts, and no doubt they enjoy the sport, though their only weapon is the cross-bow, pitted against the magazine rifle. However, even the cross-bow skilfully handled is a match for a rifle fired at haphazard in the jungle, and I was told on good authority that in a recent attack on a Lissu village, the Chinese soldiers had fired four hundred rounds without hitting a man.

Chinese activity in the Upper Salween valley dates from 1910, when two German explorers, who had attempted to penetrate into the unknown part of the valley north of Mr Lytton's explorations, were murdered by the Lissus. The rights and wrongs of this business will of course never be known, but without casting aspersions on two brave men, it can scarcely be doubted that they were themselves partly responsible for their fate, if only because they had with them an Indian cook. If there is one hard and fast rule which may profitably be observed when travelling amongst the Tibetans and tribesmen, it is this : travel with the people of the country and don't take outside Asiatics as servants. Their presence may or may not be resented as such, but it is quite certain that sooner or later they will get into trouble with the natives. Major Davies, who was accompanied by Sepoys, had experience of this when he tried to get into Tibet. Lieutenant Clarke, who had with him an Indian servant, got into serious trouble with the Mohammedan Chinese of Kansu, and lastly the German explorers added one more to the list of disasters which I firmly believe may be traced at least indirectly to this cause.

The authorities of Yunnan were goaded on to take action in this matter largely owing to the efforts of Consul Rose in T'eng-yueh, and an expedition was sent up the Salween valley to punish the tribesmen. A village was burnt, several men on both sides were killed, and some prisoners were brought down who, being apparently recognised by the Indian cook, or found to possess property of the murdered men, were beheaded. They may or may not have been the right men; probably they were not, but the Chinese practice is to punish the crime without reference to the criminal, so that everybody was satisfied.

After this, the Chinese began to realise that it would be best to adopt a more active policy in the Salween valley, and the Lissu fastness was consequently attacked both from north and south, garrisons, as already described, being established at convenient points to contain them. The Lutzu, however, are a peaceful tribe, who have, I believe, given no trouble.

In the afternoon the rain abated, though it continued to drizzle, and we began the ascent of the last ridge, passing through a belt of dense jungle at the summit, where masses of a most beautiful and sweet-scented orchid, the name of which I have yet to discover, were in flower. Presently we emerged on to the open hill-side again, and at last caught sight of the great Salween river winding in close coils through its deeply-eroded valley.

Bracken and grass clothed the hill-side; here and there were pine trees and low bushes of rhododendron and *Pteris*, but the undergrowth had been fired in many places previous to the rains and there were not many flowers in bloom, though I noticed several orchids including *Spiranthes*, one or two Liliaceae, and the tall white *Anemone Japonica*. The path was very steep and we descended at a great pace to a small hut about half-way down the mountain, where we stopped for some refreshment. Here I saw one of the most beautiful girls I have ever come across, a graceful lustrous-eyed creature, with warm sunburnt complexion. One meets such beauties sometimes amongst the Lutzu, more often amongst the Moso or the Tibetan tribes, and instinctively one wonders where they come from, for they possess no obvious Mongolian feature, neither prominent cheek-bones, nor almond-shaped eyes, and scarcely even the distinctive colouring, for the complexion may be so light as to resemble that of the European. Their straight black hair and black eyes alone betray them.

W. T.

6

Here we were given wine to drink, which was not unlike Chinese wine, a sort of crude wood-spirit. Meanwhile the girls sat on the floor talking, and slowly made string by twisting hemp fibre, bobbins of which they carried about in their bags, while the men smoked and chatted and drank their thick maize wine, which is to them much what a cigarette is to an Englishman. Presently two of the girls, sisters no doubt, sitting side by side, put their heads together and quaffed from the same bamboo cup, a proceeding which elicited much mirth. I saw this performance on a subsequent occasion, also by members of the fair sex, and again there was ill-concealed mirth amongst the elders, so that I was inclined to attach some unknown significance to it.

After half-an-hour's rest we descended to the valley, reaching the scattered village of Cho-la just as the rain set in heavily once more. The huts are built a hundred feet or so above the river on a gently-sloping platform, which is evidently an old river terrace, partly built up during the flood season and partly the result of tributary torrents spreading out their loads of detritus as they debouch from the mountains. In the rainy region these terraces form a continuous shelf between the outermost range and the river, averaging a hundred yards or more in breadth on the concave sides of the bends; but as soon as we enter the arid region, the river saws its way vertically downwards, and in spite of the windings there is nowhere any hint of a platform other than an alluvial cove. Everywhere these terraces are cultivated; rice, wheat, and maize being grown in summer and buckwheat in the autumn.

The house we put up in was large and roomy for a Lutzu hut, but I had a smaller building to myself, a tiny shanty used as a chapel for the Penates and a storehouse for maize cobs. My bed was put up alongside the altar, and I slept in the shadow of grotesque gods, while all night long the pitiless rain rattled on the shingle roof. I slumbered unconsciously through it all.

Plate XVII

Crossing the Salween at Cho-la in dug-out canoes

CHAPTER VII

THROUGH THE LUTZU COUNTRY TO MEN-KONG

On the following morning the rain gradually abated, and not only did it cease entirely before mid-day, but the cloud canopy began at last to disperse; it had rained continuously for the last fifty-six hours, and I was becoming rather depressed.

We now crossed the river in dug-outs, the Lutzu paddling us across with large square-bladed paddles, though there was a fine current running, the river being forty or fifty yards broad, considerably bigger and swifter than the Mekong. These dug-out canoes closely resemble those used by the Malays, but the dug-out is of necessity almost exactly the same everywhere, and was so widely distributed over the world at such a remote period that it probably originated in a number of distinct centres, and its presence in this place or that is not necessarily any evidence of connection between the two peoples.

The high rocky banks of the river are composed of purple slates tilted nearly vertically, and slate instead of wood was frequently used to tile the huts. Above the alluvial platform, however, bare scarps of limestone were frequently conspicuous. The same slaty rocks appear on the Mekong much further south, where they are also tilted vertically, or nearly so. The river makes remarkably close S-shaped curves, but is not here interrupted by serious rapids, and we saw several Lutzu in their canoes fishing. Close inshore, nets attached to stakes are occasionally set up, but most of the fishing is carried on from canoes, a V-shaped net suspended between two long bamboos being plunged under water; the legs of the V are then widely opened and the net lifted to the surface.

We lunched at the village of Ze-tou, a few miles up the river.   The alluvial platform was now confined to the right bank, but there was not much cultivation, beans and barley being the chief crops.   Deep gullies cleft the range behind, affording glimpses of high forest, and in the shadow of these cliffs the vegetation was extremely rich, a considerable variety of terrestrial orchids being particularly noticeable, also ferns, irises, campanulas, and a single plant of *Lilium giganteum*, the only one I saw in flower.   It was six feet high, but would grow taller yet, with half-a-dozen long white trumpet-shaped flowers two or three inches across. On the drier slopes were pine trees, ferns of the genus *Pteris*, and a lycopodium, but in the cultivated parts we walked between hedgerows of large St John's Wort and yellow roses, lined with blue irises.   At last we came to a deep ravine, dominated on the other side by a high bluff, on which T'sam-p'u-t'ong is built; and at the top there stood waiting for us a solitary figure, clearly silhouetted against the sky.   It was a Chinese soldier.

With commendable and surprising brevity, he asked just sufficient questions to identify me and no more, but I had told Gan-ton to be quite vague as to our ultimate destination, not being quite certain of it myself.

T'sam-p'u-t'ong is a small village built on a steep slope, backed by a fine limestone cliff.   The wooden huts were as usual scattered, the monastery dismantled and falling to pieces, the yamen unpretentious, being indeed part of the derelict monastery.   The thirty priests were apparently dispersed ; presumably they are or were Tibetans, not Lutzu.   The population of the village consisted of perhaps thirty families, Lutzu, Tibetan, and a few Chinese, with an official and a garrison of ten soldiers.

The official himself accosted me while my men were vainly assaulting the locked doors of the lamasery, and introduced himself in an off-hand manner ; whereupon we repaired to the yamen to drink tea.   After a little talk he kindly offered to open part of the lamasery for me, but it was in a most dismal state of disrepair, though there were some excellent wall paintings.   The official, having failed to discover by what route I intended to return to A-tun-tsi, remarked urbanely before taking leave, that of course I

Plate XVIII

Lutzu woman of Cho-la

A Lutzu woman of Cho-la
carrying a bamboo water-tub

Lutzu cross-bow man of Cho-la

would not go into Tsa-rüng "as the road was very bad"—
the usual formula of every official who politely deters one
from making a journey—and he would be happy to provide
me with an escort to see me safely back.

Of course I acquiesced, so next morning, we continued
our journey up the valley with an escort of two soldiers.

Looking westwards, I noticed for the first time that a
big valley here slits open the Salween-Irrawaddy divide, and
subsequently I was told that by following up this valley,
the Irrawaddy might be reached in four days, though the
journey was described as one of great difficulty and danger.
It was some months later that I met in A-tun-tsi the former
official of that place, a man named Hsia-fu, who had himself two years previously made the journey across from the
Salween to the borders of India by this very route, and
had subsequently been dismissed from office. At the same
time I met 'Joseph,' who in 1895 had guided Prince
Henri from Tali-fu to India; but their route lay to the
south of T'sam-p'u-t'ong. Both men were disposed to
make the most of their hardships, which were certainly
severe, since the journeys were undertaken during the
rainy season, when mosquitoes and leeches were at their
worst, every river a roaring torrent, and the vegetation so
dense that it made progress terribly slow.

Some of my porters having returned to Tsu-chung,
three of the loads were now carried from village to village
by relays of Lutzu, usually women, frequently dwarfish or
deformed people, who were possibly slaves.

Immediately above T'sam-p'u-t'ong we entered a magnificent limestone gorge, and here the results of the heavy
rainfall in the Salween valley were fully displayed.

The narrow path through the gorge was shaded by a
great variety of tall, straight-limbed trees, including several
Asclepiadaceae and some species of *Ficus* with flowers
borne directly on the old wood, their straight unbranched
trunks frequently supported by feebly-developed plank
buttresses. A great variety of lianas such as *Rhaphido-
phora*, and creepers, made the jungle still more dense, and
amongst an array of epiphytes I noted, besides numerous
orchids, a fern which I could not distinguish from the
tropical *Asplenium nidus-avis*, though I have never before

seen it so far north.   Amongst the undergrowth was a big erect selaginella (perhaps *S. grandis*), and here many of the orchids were in flower, amongst them being a *Dendrobium* with long pendent sprays of orange flowers.   Yet, in spite of the heavy rains in the Salween valley, it is only the deep gorges and gullies which are filled with this dense monsoon jungle, the fully exposed mountain sides being covered with bracken, grass, scattered pine trees and low scrub.

The obvious inference is that rain alone does not satisfy the monsoon forest, for it cannot stand full exposure during the dry winter, the reason being that the only source of water, namely the heavy dews, are rapidly evaporated in the hot sunshine.   Hence on the exposed slopes of the broad open valley there are found chiefly Conifers (*Pinus*) and semi-xerophytes, for most deciduous-leaved trees of the north temperate forests require water at all seasons of the year.   Similarly, higher up on these exposed slopes are forests of spruce only, while in the deep narrow valleys are alder, birch, maple, and many others.

Just above the gorge is the village of Lang-chi, and here a big torrent pours itself into the Salween from the west, the latter tumbling several feet over an enormous accumulation of boulders, with a noise like thunder.   It was the first big rapid we had seen on the Salween.

A little further up on the other side is the village of Wu-li, and now we came upon two rope bridges, by the second of which we crossed to the left bank.   Here I was interested to observe that the Lutzu when crossing the rope bridge always carries with him, fastened by a loop to his wrist, a bamboo tube full of water which he tilts just in front of the slider.   This doubtless reduces the friction and, by keeping down the temperature, lessens the wear and tear on the rope.

Seeing the rope bridge here in the dense jungle, set me thinking of its origin.   I surmise that it is the practical application of the natural liana bridges probably to be found stretching across the headwaters of the Upper Irrawaddy, as they certainly do in jungles elsewhere.   If so, this would lend more colour to the suggestion that the Lutzu are immigrants to the Salween from the west, the idea of the rope bridge having been copied by the Tibetans, and carried right away eastwards.

In the afternoon there was a very perfect sun halo, a phenomenon which I had seen several times in the Mekong valley, usually followed by rain. It is perhaps caused by water vapour drifting across the hot valley from one mountain range to the next, under the influence of the south-westerly wind. I never saw this halo in the arid regions further north. There was an abundance of insect life in the valley, butterflies and beetles being particularly numerous, though I noticed no uncommon genera. Brilliantly coloured *Hemiptera* were another feature, besides long green stick-like insects (Mantises) and many lizards. Birds however were few, and I saw no other animals.

Later in the afternoon, at a point where the Salween emerged from another deep and narrow gorge, we turned up a valley to the east and ascended to the village of K'un-a-t'ong.

Here there was a small chapel, and I called on Père Bernadier, who had spent forty years of his life on the Tibetan frontier in the service of the Catholic Church, during which time he had only once returned to France. I found him a most interesting man, and spent a very pleasant evening in his company. He told me that, with the exception of an American journalist named Nicolls, all the travellers in this region had been Frenchmen, beginning in 1895 with Prince Henri, and ending only the year before with M. Bacot, and I learnt something of what the Catholic priests and other travellers had done. It is curious to reflect now, that at this very time two other Englishmen, Captain Bailey and Mr Edgar, were at Men-kong, whither I had decided to push on.

The French mission on the Tibetan frontier has been subject to many vicissitudes, and its centres have again and again been moved, so I doubt whether any lasting impression has anywhere been made in the Salween valley. Certainly I must confess I saw no obvious results of it amongst either the Lutzu or Tibetans, for I cannot honestly ascribe their friendly attitude to the good offices of the Catholic Church. The latter at least are hospitable or hostile according to the whim of the moment, hospitality prevailing, and the former, I believe, are never anything else but friendly. Personally I think we owe more to these bold priests for their additions to our geographical and scientific knowledge of the country than to their efforts at proselytizing the natives.

I had now, as already stated, decided to push on up the Salween valley as far as Men-kong, the obscure capital of the province of Tsa-rüng, but first it was necessary to get rid of my escort, for I did not feel comfortable at the prospect of travelling either amongst the Lutzu or the Tibetans with Chinese soldiers, who are apt to be over-bearing. I had found my escorts, even on the main roads of Yunnan, rather a nuisance, dislocating the traffic and overawing the people. There, however, it did not much matter, and an official escort always looked well, for it showed the peasants plainly that my presence in their country was acknowledged by the officials, and if there is one thing the Chinese peasant reveres after his ancestors, it is a mandarin or anyone under his protection, and escorts are the hall-mark of the yamen. But in Tibet it is different, for the Chinese 'brave' is not, except on the main road, the familiar object there that he is elsewhere; and being intolerant of 'wild men,' as he calls them, he is apt to get into trouble with these bold independent spirits. For myself I had no fear, as I was travelling with the people of the country, and was not likely to do anything foolish.

Consequently I told the two soldiers who had escorted me from T'sam-p'u-t'ong that I was staying here a week to collect plants, and that they had better return. I then gave them a small present as is customary, and they started gaily back the same evening, no doubt to be soundly rated for letting me out of their sight. In the night I changed my mind and decided to go straight on next day.

K'un-a-t'ong boasts a population of fifteen or twenty families, with more scattered at intervals up the valley. It is the highest point on the Salween where rice is cultivated, and when next we saw that river a marvellous change had taken place in the character of the country.

Unfortunately it was now necessary to make a two days detour back into the mountains, on account of the impassable gorges through which the river flowed. A narrow path, said to traverse these gorges, was now under water, though exposed during winter ; so narrow was it, however, that loaded porters could not at any time negotiate it. Could one ascend straight up the valley there can be no doubt that from the jungle of the K'un-a-t'ong gorges to the beginning of the arid regions at Saung-ta would be no more than a day's march.

The chief at K'un-a-t'ong, a tall broad-chested man with a strong pleasant face, obtained three porters for me, and the villagers were well satisfied when I paid over their wages to him before we set out. This man, who was richly dressed and carried a long spear, was a fitting counterpart to the beautiful girl I had seen previously ; on the other hand, my three temporary porters were dwarfish and criminal-looking, bearing little resemblance facially to the ordinary Lutzu type. What struck me more than anything else, however, was that one of them had a distinctly negroid type of countenance, with thick lips and flattened bridgeless nose, and instinctively I recalled a Tibetan woman in A-tun-tsi who had precisely these characteristics developed in the same unmistakable way. There can indeed be no question but that amongst some of these Tibetan and other tribes a negroid type does occasionally crop out, but for the present, at least, I forbear to speculate on its origin and significance. All my Lutzu porters, both men and women, carried their loads by means of a strap made of twisted bamboo strands passing round the forehead, a method commonly adopted by dwarfish races, as already remarked, and especially by jungle tribes, being common, for example, in Borneo ; but it is never adopted by the Tibetans, who carry their loads on their backs by means of shoulder-straps. The Lutzu are not indeed dwarfs, but they are distinctly short in stature, and we have here another argument in favour of their jungle origin.

We had scarcely started up the valley when the rain began again and continued all day. Presently we entered the forest, where many of the big trees were covered with epiphytic ferns and orchids and draped with moss. Amongst the undergrowth dense patches of an Impatiens and the

pretty little Enchanter's nightshade (*Circaea*) were conspicuous, besides the Oxalis already referred to.   Higher
up we came to towering limestone walls with bunches of
flowers hanging from every crack and cranny—the small
mottled *Cypripedium guttatum*, *Primula sinensis*, saxifrages,
Epimedium, begonias and other pretty things.

Camp was pitched under one of these big cliffs, where
the stream dashed through a deep gorge, and while the
men settled themselves beneath a protecting ledge of rock
and built up a big fire, my tent was erected amidst the
soaked undergrowth.   I was wet through, and for once
thoroughly tired out by the long climb through the forest.
For a long time we sat round the fire chatting, Gan-ton
and Kin happily being in the best of spirits.   The latter
had borrowed a Tibetan cloak, and walked into my tent in
this guise to ask me what I thought of him, somewhat to
my amusement, for I failed to recognise him.

As the men, one by one, dropped off to sleep, I left the
warm little nest under the cliff and crept into my cheerless
tent beneath the dripping trees, leaving them all curled up
round the fire like cats.   It grew colder and colder in the
night, pouring with rain throughout, and I awoke to hear
Ah-poh barking furiously; next morning Gan-ton told me
that it was because he had seen wild animals in the forest,
deer perhaps, or bears.

June 16th was our longest and hardest day, and still
the rain continued.   I got up feeling very unwell, and the
long slippery climb up an awful path made me feel worse
than ever; but the men stuck splendidly to their work.
An ascent of about 4000 feet through rich forest brought
us to the pass over the spur, but when we wanted to stop
for lunch below the summit, the men were unable to light
a fire, so we had to go on.   Every Tibetan or tribesman
when crossing the mountains carries under his cloak a few
chips of resinous spruce wood, so that with his flint and
steel and a little dry grass for tinder, he has all the necessaries for starting a fire.   To secure this resinous wood, a
big tree is selected and burnt on one side, a process which
seems to attract abundance of resin to the wound.   The
charred bark is subsequently sliced away, and pieces of
wood cut from the trunk.   Along the forest paths these

blazed trees, with quantities of resin exuding from the wound, are a common sight, and the Tibetans, when they meet with one, hack out with their swords a fresh supply for their torches, or for starting a fire.

I found that there are definite camping and halting-places on all these trails, sometimes under a cliff, perhaps, such as the place we had occupied on the previous night, or under a big tree, the blackened trunk of which has been hollowed out by a long succession of camp fires.

The torrent we were following up to its source was frequently interrupted by small waterfalls, due apparently to the presence of harder shale strata amongst the limestone rocks; but in the stream bed were blocks of olivine, red porphyry, and other altered igneous rocks. A dense herbaceous vegetation five to six feet high, composed mainly of Ranunculaceae such as aconites, columbines, and larkspurs, with Polygonum and saxifrage, lined the sides of the forest, in which arborescent Ericaceae, especially rhododendrons, were conspicuous, with spruce and larch predominating towards the summit.

Crossing the spur at 12,000–13,000 feet we turned north, and during the descent gradually worked round to the west again. *Primula sonchifolia* was still in flower up here, and lower down were masses of the shrubby *Paeonia lutea* and the two species of *Cypripedium* already referred to several times. On the limestone cliffs I noted for the first time bunches of the delicate violet-flowered *Isopyrum grandiflora* (Ranunculaceae) of which I afterwards found quantities on the limestone cliffs above A-tun-tsi.

The descent to the Salween was a long and trying affair, for the mountain side was extremely steep and slippery, and my toes were sore for days afterwards from being pressed so hard into my wet boots. Gradually the high rain forest gave place to pine-clad slopes, and finally we found ourselves in a deep ravine with scattered huts and patches of cultivation above. Masses of orchids were in flower on the shale rocks, but as we approached the Salween everything seemed to shrivel up, and there were left only low-growing leguminous shrubs, and myriads of globular selaginellas. At dusk we got down into the Salween valley and approached the village of Saung-ta, when suddenly a number of men came out from the houses and advancing towards us, made us welcome in the most friendly manner, bowing and smiling to me, and exchanging conversation with the men. Thus we were escorted into the village by the chief and his friends, and were quartered in the best house available.

Saung-ta contains about twenty-five houses built along the edge of the river bank. Behind it is a very narrow cultivated platform, and cultivation extends up the lower slopes of the mountains for some distance. Just below the village, the river narrows and silently enters the gorges, but here it is divided by an island of shingle and chatters merrily by. There are a few houses on the opposite bank also, the Lutzu crossing backwards and forwards in their canoes, several of which were drawn up on the beach. The houses are built, not with gable ends and slat roofs, but in the Tibetan style, with flat mud roofs which are reached through a square aperture in the centre; an open shed usually covers one side of the roof, forming the second story, and there may be two or three rooms down below, instead of the one large room common to the Lutzu huts at Cho-la. A few miles higher up the Salween, the Lutzu finally give place to the Tibetans, just as on the Mekong the Moso and Lissu tribes do, and it is obviously their influence that we see at work here.

Tibet, it must be remembered, is largely a region of deserts and semi-deserts, using the term in its widest significance to include regions rendered more or less devoid of vegetation by salt steppes, by wind, or by lack of rain, and towards China at least the limit of desert conditions marks the limit of the race. Perhaps it is just these highly abnormal conditions which render the agricultural Chinese immigrants powerless to cope with the situation, and cause their gradual absorption by the thoroughly acclimatised Tibetans.

At Cho-la and T'sam-p'u-t'ong we have, if not the original Lutzu architecture, at least remnants of it, but this type of log hut with slat roof is, as a matter of fact, built by many tribes, being indeed almost universal in the Mekong and Salween valleys till we get down south amongst the Shans.

Plate XIX

A Lutzu girl of Cho-la weaving hemp cloth

Lutzu of Saung-ta in the arid region of the Salween valley

The most extraordinary change observable in the Lutzu tribe, however, was not in their houses, but in their persons. Their tangled hair, cut in a short fringe across the forehead, hung matted over back and shoulders; the men and women were in Tibetan dress though they wore no boots and little jewellery, but the girls wore the skins of goats or precipice sheep sewn together into a sort of sleeveless overall with the hair inside, and the children went about naked. They were, without exception, the filthiest people I have ever come across, and had not the excuse of climate for their marked aversion to water. Their complexions were darker, and their features, instead of being more Mongolian, were, if anything, more negritoid than those of their relations down the river. Their vessels are of bamboo and pottery, they weave hemp, till the fields, and fish. Barley, maize, and hemp are grown, besides pomegranates and prickly pears. Cattle and chickens were the only domestic animals I noticed, and we were able to buy eggs, milk, and butter.

Wooden rosaries are commonly worn by the men, metal bangles and ear-rings by the women ; the men frequently wear one ear-ring only, in the left ear. There were a few priests in the village, both men and women, with closely shaven head, but we saw no religious ceremonies performed. They had a curious habit of smelling anything new at the very outset of their examination of it—my clothes, a piece of canvas, even a piece of silver paper, were all smelt critically, which suggests at once that they must possess a keen and discriminating sense of smell. The general uncouthness of these people must be ascribed to a variety of causes, amongst which the impossibility of cultivating rice, their isolation from all but the scattered Tibetan settlements higher up the river, and the change of climate, are no doubt important. But undoubtedly the chief cause of divergence is a negative one, namely the absence of Chinese influence.

The Chinaman unconsciously influences those tribes with whom he comes in contact—excepting always the Tibetans—so that they gradually lay aside their more barbarous habits. The Lutzus on the borders of the arid region may not perhaps precisely resemble the original tribe who came, as I believe, from the jungles in the west, but they are likely to be far less changed than the Lutzu of T'sam-p'u-t'ong. The former we may call 'black' Lutzu, not entirely with a detergent significance, to distinguish them from their more refined relatives to the south.

My tent was pitched on the flat roof of one of the houses. The rain had ceased, and a warm breeze blew up the valley; a few stars shone out through rifts in the clouds, and tired as I was after eleven hours climbing in the rain, I turned in and slept like the dead. The sun was high in the heavens when I awoke and heard below me the river rattling over the end of the shingle bank.

For the remainder of our journey up the Salween valley the weather was glorious, yet trying. About mid-day a scorching wind began to blow, increasing in violence till the middle of the afternoon, and gradually dying out after sunset. The deep U-shaped valley became a V-shaped gorge like that of the Mekong, the cliffs became more and more bare, the dryness of the atmosphere steadily increased till it became intolerable, and day after day the sun glared in the ribbon of blue sky which faithfully followed the windings of the deep valley. But on the mountains east and west poured the everlasting rain. It was an exact repetition of the Yang-tze and Mekong, and further, in the case of the Salween and Mekong at least, the arid regions begin in exactly the same latitude.

The jungles of the Salween, as we have seen, give place abruptly to the arid gorges above T'sam-p'u-t'ong, and it would be possible to pass in a day from a region where it rains practically every day for six months in the year (it is useless to hazard what the rainfall might be) to a strip of country about two miles wide where the annual rainfall certainly does not exceed ten inches, and may be substantially less.

The Salween-Irrawaddy divide stretching north and south with its high snow-clad peaks above T'sam-p'u-t'ong checks the rain-bearing winds from the south-west, thus acting as an efficient rain-screen to the valley further north. Any clouds which succeed in crossing this range throw down their moisture, not in the deep and narrow valley, but on the now greatly elevated Mekong-Salween water-shed, which, from an average altitude of 15,000–17,000 feet opposite T'sam-p'u-t'ong rises suddenly to the terrific peaks of K'a-gur-pu and its northward extension.

The Mekong-Salween divide then, and particularly the great pyramid of K'a-gur-pu, acts as a second rain-screen, and with such effect that it not only cuts off almost the entire rain-supply from the Mekong valley, but also to a very large extent from the next range to the east, the Mekong-Yang-tze divide, thus elevating its snow-line two or three thousand feet. The effect is still further enhanced by the fact that two big mountains rise above the snow-line on this range also, opposite K'a-gur-pu, and when the winds reach the Mekong-Yang-tze watershed, almost stripped of their moisture, what little remains is condensed by these high summits. This unequal distribution of rainfall on the ranges east and west of the Mekong has a considerable influence on their physical features, on the composition and distribution of their flora, and other points to be discussed later.

At mid-day we reached the last Lutzu village, called Laung-pa, and after lunch embarked in a dug-out for a voyage through some more gorges; and an interesting voyage it proved. This dug-out was barely 24 feet in length, about 18 inches in the beam amidships, and the same in extreme depth, but there were no less than twelve of us on board besides the dog and the luggage. We squatted on our haunches in single file, there being five paddle men forward and a single steersman aft. Our gunwale was almost awash, and considering the water we had to go through, I frequently thought we must capsize. The men hugged the walls of the gorge when they could, pushing against the rocks with their paddles, then darting across the river to avoid a rapid and catch the back-current. Whenever shingle islands or a shore line appeared, they got out and tracked, hauling the canoe through a lot of rough water with their bamboo rope. Thus we covered several miles till we came to the first Tibetan village and finally disembarked. The last I saw of the Lutzu they were drifting down stream in the canoe, sweeping their V-shaped net under the water. There was no hint of rain now; it was a glorious evening, and a scorching wind raged up through the gorges, which grew more and more

wild and lifeless. We did not reach our destination till
nine o'clock, but it was a brilliant star-lit night, delightfully
warm.

La-kor-ah consists of three huts and a temple, in the
shadow of which the tent was pitched. It was a sanctified
spot, and dozens of prayer-flags made it look larger than
it really was, while on each side of the temple stood a row
of big leather prayer-drums, much the worse for wear,
containing probably hundreds of yards of the everlasting
prayer. Each passer-by set them revolving one by one,
the rusty spindles groaning fearfully in their sockets.
Immediately below, a grey glacier torrent came booming
through a deep sword-cut in the mountains and sweeping
down into the mighty Salween was instantly engulfed in
a surge of yellow waters. Up this narrow rift lay the
pilgrims' road to sacred Doker-la. Here Gan-ton learnt
from the residents that the French traveller M. Bacot was
at Men-kong with a large number of mules, and I looked
forward to meeting him; but in this country one rarely
hears the truth of a story the first time of asking.

Next day's march through a terribly arid and totally
uninhabited stretch of the valley was a tiring one, though
the track was surprisingly good. The river swept in huge
S-shaped curves round colossal buttresses, smashed its way
through deep gorges, and roared over the boulders. Immense screes, sometimes smoking with the dust of falling
rocks, rose bare and lifeless on either hand, and the ceaseless
scorching wind, which seemed to suck the vitality from
everything, blew throughout the day with ever-increasing
violence. Once in crossing a scree, I narrowly escaped
being hurled into the river by a small avalanche, but
hearing a peculiar noise I glanced upwards in time to see
a cloud of rocks whizzing through the air, whereupon I
turned and ran, reaching safety just as they hummed past.

Under that incandescent sky, stretched like a tongue
of fire up the valley, the place became an oven, but the
mountains to east and west were as usual buried in cloud.
However from the village of Chia-na we watched the sun
sink in a wild blaze of colour behind Men-kong, now only
a few miles distant. Above Chia-na a narrow stony valley
to the east led to another pass across the watershed. The

Plate XX

The Monastery, Men-kong

Men-kong, capital of Tsa-rüng, S.E. Tibet, Salween Valley

lama caravans which pass through A-tun-tsi go this way
to Lhasa, joining the main road again at Chiamdo, to the
north of Men-kong.

Passing between boulders of granite amongst which
grew masses of Opuntia now in flower, we reached the
capital of Tsa-rüng before mid-day, and leaving the men
to fix the camp, Gan-ton and I crossed to the right bank
by the rope bridge, and climbed the cliff.

Men-kong is built on an alluvial fan washed down from
the mountains by two converging torrents, and ending
abruptly in a bluff some six or seven hundred feet high, the
Salween—here almost continuously interrupted by rapids—
flowing in a deep trench below. Scattered down the slope
are the big two-storied 'manor' houses, standing amongst
fields of waving corn and shaded by magnificent walnut
trees; the contrast between the golden barley and the olive
green foliage, from amongst which the white houses peeped
out here and there, was charming.

Up on the hill side, almost under the shadow of the
forest, stands the ancient monastery, its splendour dimmed
by the ravages of time, unheeded by the priests and people,
and westwards the neglected road winds away over the
mountains to the plains of Assam. The stone-paved courtyard facing the temple is empty now, and the heavy doors
of the temple itself are locked. Only the wind sighs gently
through the sombre arbor vitae trees which spring up here
and there amongst the courts and little wooden houses,
some of them decked with flowers where the priests
reside, and the ragged prayer-flags flutter merrily. Even
as I stroll through the deserted court-yard, its red and
white walls almost flashing in the brilliant June sunshine,
I hear the rise and fall of the flails, and the chorus of
Tibetans singing perhaps their harvest song.

The population of Men-kong, including adjacent settlements across the river, is given as seventy families, and the
monastery contains about a hundred priests. There is also
a garrison of fifty Chinese soldiers from Ssü-chuan, and
after visiting the monastery I paid my respects to the
military official, a very pleasant young fellow. He did not
seem in the least surprised to see me; rather did he
look as though he expected an Englishman to stroll into

W. T.

Men-kong every day, and I soon learned the reason. I had arrived just too late to meet two Englishmen (afterwards identified as Captain Bailey and Mr Edgar, the English missionary at Batang).

It was bitter luck to have missed them! How splendid if the three of us had been able to foregather there, and compare notes of our journeys which had only overlapped for the eight or ten miles from Chia-na! M. Bacot, who in the previous year had been, I believe, the first European other than the Catholic priests to visit Men-kong, was not there at all; the mistake as to who the mysterious European was in front of us arose from the fact that Captain Bailey and M. Bacot (who had been accompanied by my guide Gan-ton) possess the same Chinese name.

From the military official I learned that India might be reached in three weeks, and that eight days journey to the west was the village of Chi-göng, where another Chinese post had been established, and only a day's journey from Chi-göng was Rima, also Chinese. These places are of course within the basin of the Upper Irrawaddy, so that it will now be impossible to mark the limits of Upper Burma by means of a primary physical barrier such as the Irrawaddy watershed. One other feature of Men-kong I can only mention here. Mr Edgar, who had slept in the village, told me afterwards that he had seen there a number of slaves belonging to a dwarf tribe which he could not place amongst any known people. This is certainly a discovery of remarkable interest, and we may hope to hear something about these dwarfs from Mr Edgar himself before long.

We returned to camp for supper after an interesting day's sight-seeing, but I regretted that I could not start westwards in the wake of Captain Bailey who had, I was told, set out for India only two days previously.

Plate XXI

Priests' houses, Men-kong monastery   Mani pyramids: the peak of K·a-gur-pu
in the background

CHAPTER VIII

DOKER-LA—THE SACRED MOUNTAIN

In the evening I climbed down the steep river bank, and standing on the rocks had a much needed bath, though it was of course impossible to take a dip in the river. In the mountains the cold drenching rain discourages one from stripping even in summer, and in the arid gorges opportunities for bathing are few.

Stores were now running very short, a matter which, without being in the least serious, greatly affected my comfort, for I was altogether out of such luxuries as coffee, tea, biscuits (in lieu of bread), butter, jam, milk, and bacon, though I had still a few tins of meat for the journey across the mountains and was able to purchase butter and eggs at Men-kong.

On June 20 we started back, reaching Chia-na before twelve o'clock, but further the men would not go that day as they were bent on cooking a goat they had just purchased, and I had perforce to give in. Moreover Kin, who had been ailing for a day or two, seemed very unwell.

On the following day we retraced our steps to La-kor-ah, halting at mid-day in a little bay where the limestone rocks receded from the river and a spring of clear and slightly warm water gushed out, filling an artificial bath and giving rise to an oasis of delicate ferns and moss in a grey-white world of rock. The life-giving stream was an obvious camping place, and the devout Tibetans, doubtless awed at the wonderful phenomenon of pure water welling out of the solid cliffs in this wilderness, had made obeisance to it by carving the sacred prayer and numerous other inscriptions, some of them of considerable length and beautifully executed, on the surrounding rocks; and amidst these manifestations of piety I bathed.

Several pilgrims from Doker-la passed during the morning, and for some reason Ah-poh went nearly mad with rage at sight of them.

We reached La-kor-ah on the evening of June 21 in glorious weather; and next day Kin was seriously ill. Curious purple patches had come out on his chest, arms, and legs, and I was baffled to know what was the matter with him; but as he complained chiefly of vomiting, I thought it likely enough that he had caught a slight chill, and having dosed him with brandy, I put a mustard plaster on his chest and made him wear a cholera belt I fortunately had with me. To stop at La-kor-ah, where we could get nothing, was useless, and I asked him to make every endeavour to travel.

The military official at Men-kong had kindly provided me with a pony for the return journey to La-kor-ah, but unfortunately I had sent the beast back on the previous evening, and here we could secure no riding animal for our invalid. However by feeding him on milk and soup, and administering frequent doses of brandy, we had the satisfaction of seeing him struggle pluckily through a hard day, though he looked ghastly enough.

From La-kor-ah we ascended the stream by the pilgrims' road to Aben, a mean little village of about forty huts, where I was well received. All the way up through the gorges we found the cliffs decorated with prayers, pictures of Buddha, many of them coloured, and other artistic designs, some of the longest paragraphs being exquisitely carved in minute characters, but the familiar *mani* prayer in bold characters usually occupied the more prominent places. What monumental patience was exhibited here! for though the surface of the rock was smooth it was rarely flat, and whole paragraphs had been written inside natural cavities extremely difficult of access, even for a hand.

At Aben we obtained a ponderous yak to transport some of the baggage over the mountains, but riding animals could not be procured here either, though doubtless they existed. A little to the south-east lay the village of Boonga where the Catholic priests were persecuted so terribly many years ago.

For a week we had tramped beneath a flaming sky between brazen valley-walls raked by a wind as from the mouth of Hell. Abruptly the weather changed again and ere we had ascended far into the pine and oak forests above Aben, surmounting a wonderful limestone gorge, the rain was once more drenching us through and through. Finally we dropped down beside the torrent again and pitched camp under a huge cliff, but there was no room to put up the tent and we huddled as closely as possible under the rock wall to avoid the rain, my bed being set up in the open. Soup and tinned meat had now to be prepared for Kin, who required careful nursing, and I had my hands full for the evening.

Looking due east up the valley we occasionally caught a glimpse of the K'a-gur-pu glaciers, from which this big torrent obviously rises, though hitherto the main stream has been drawn on the maps as rising in the south-east from Doker-la itself. The reason for the mistake is no doubt partly due to the confusion introduced by Prince Henri d'Orléans in calling the snowy range Doker-la, and partly to neglect of the snowy range altogether, so that, in order to account for the size of the torrent at La-kor-ah, its source has had to be placed some distance to the south.

Leaving the main valley next morning we climbed a spur and found at the summit the usual bundles of bamboo wands decorated with strips of paper and rags, but what was more odd still, there were ranged on the ground, on the rocks, in the grass, everywhere, rows and rows of empty *tsamba* bowls as votive offerings. There must have been scores of them.

Bearing away to the south-east so as to avoid the snow peaks, we pitched camp once more in the forest at an altitude of about 10,000 feet, with the prospect of crossing Doker-la on the following day.

I was now completely out of stores with the exception of some soups which I kept for Kin, who happily was feeling much better. I was still giving him brandy night and morning—though we were nearly out of that too—and feeding him chiefly on lightly-boiled eggs, of which we had brought a moderate supply from Aben. It seemed to-night that I was destined to sup off *tsamba*, eggs, and brick tea, not a very satisfying meal at this stage of the journey, but Gan-ton came to the rescue with some toadstools and bamboo shoots, the latter being roasted by the simple expedient of throwing them into the fire and leaving them till, on stripping off the outer burnt leaves, the inside was found to be soft and succulent. As to the toadstools, a sort of morrell, I knew nothing about them and had to trust to Gan-ton's empirical knowledge of jungle produce; but he ate some himself, so that, unless his colour carried with it immunity from the effects of vegetable poisons, I anticipated no evil results. A minute but ferocious insect like a sand-fly existed in the forest here and gave us a bad time, settling on our faces and wrists in hundreds, and biting us till they raised most painful bumps.

June 24 was our last heavy day, but I now felt very weak from lack of accustomed food, the continual change of weather, and sheer weariness, so that even the sombre and bulky yak easily kept in front of me. Gradually we emerged from the forest into a region of alpine meadow dotted with willows, and covered with beautiful flowers, such as yellow salvias, purple columbines, and masses of the red *Fritillaria Souliei*, which grew waist high. Craggy limestone cliffs rose on every hand, but the whirling mists hid the view.

Lunch was taken under the shelter of some boulders amidst patches of snow, whither I was the last to arrive, feeling very bad. Suddenly there came a hail from one of the Tibetans just outside our little shelter, and everybody scrambled excitedly for the open and climbed on top of the boulder; sacred Doker-la had become momentarily visible through the driving mists.

We had now reached the belt of alpine grass-land or turf, also represented on the Mekong-Yang-tze divide, though the rainfall is not there sufficient to give rise to a definite belt of alpine meadowland. The ascent to the pass was as usual extremely precipitous, and nothing being visible from the summit save tantalising peeps through the ever-shifting cloud veil, I was glad to descend. A driving wind beat the cold rain in our faces, baulking every attempt to start a fire and obtain a rough boiling-point reading. However, though there was now no snow on the summit, I think that the Doker-la must be a little higher than the Sie-la.

Why this mountain should be so sacred I cannot think, for there is nothing about it to stir the imagination. It is easy to understand people bowing in awe before the virgin peak of ice and snow to the north, and it is quite possible that on a clear day K'a-gur-pu would be readily seen; but Doker-la by itself, in comparison at least, looks rather mean, though the weather may have biased me.

On our journey up we had passed several small parties of supplicants returning from a visit after adding their quota of rags to the prayer-scattering bamboo wands which crowned the pass, but autumn is the proper pilgrim season, and in October hundreds of them passed through A-tun-tsi. Still I firmly believe that it is K'a-gur-pu that really inspires them, though Doker-la may well be the easiest approach to a somewhat inaccessible peak. Tibetans do not mountaineer for fun, and even sacred Kailas is worshipped from afar.

This day's journey finally shattered the belief to which I had all along clung, namely that we were going to cross the snowy range, though our turning south-east on the previous day had made this seem very problematical. A rough precipitous descent finally brought us down to a deep valley in which flowed a big torrent, obviously another glacier stream from K'a-gur-pu, which could not have been very far to the north, and we halted for tea under the shelter of an enormous granite boulder, the rock on this side being mostly granite, in strong contrast to the quartz-bearing limestone on the other side.

Another thing worth noting is that the watershed is asymmetrically placed, being considerably nearer the Mekong than the Salween, nor are there here, in the arid region towards which we were descending, any barrier ranges. In the rainy regions we have already remarked that the streams flow from their source for some distance parallel to the main rivers, so that the main spurs are often parallel to the main range for some distance. In the arid regions, however, the torrents flow directly down from the mountains, and the main spurs are always at right angles to the watershed and consequently the passage of the range is less fatiguing, though not necessarily any shorter.

We camped far down in the forest that night, and saw the stars shining overhead once more, both Kin and I feeling very much better, though dead tired.

It may be interesting at this stage to institute a comparison between the plant formations on the mountain range we had now crossed for the second time and the Mekong-Yang-tze divide, where I had been climbing in May and subsequently crossed several times.

The sequence of belts on the Mekong-Salween divide, starting from the Mekong, is roughly as follows :—(i) Grass-land and bracken, with scattered pines, rhododendrons and so on, (ii) Forest, with fir trees and giant rhododendrons, numerous shrubs, and deciduous-leaved trees. In the shady gullies this temperate rain forest descends right down to the Mekong and Salween, (iii) Forests of birch and alder with dense undergrowth, (iv) Alpine meadow, (v) Alpine grass-land or turf.

Turning now to the mountains above A-tun-tsi, we find that two of the above belts, namely birch forest and alpine meadow, are entirely wanting; that the larch becomes an important constituent of the upper forests on north-facing slopes ; and that a new element, namely the scree flora, makes its appearance.

A full consideration of these facts is beyond the scope of a story of travel however, and it will suffice to say that these differences are directly or indirectly traceable to the difference of rainfall on the two ranges of mountains ; that the composition of the forests on the two ranges differs more than the composition of the alpine flora on account of the greater importance of rain to forest land than to grass-land; that in the alpine meadow of the Mekong-Salween divide are to be found several plants, particularly Liliaceae, which are not found on the Mekong-Yang-tze divide ; and finally, that the limit of plants is substantially lower on the former than on the latter, in accordance with the lower snow-line, the extra ground available on the Mekong-Yang-tze divide being favourable for the production of new species which, so far as I could make out, are peculiar to that range.

On the following morning we reached Londre, where we learned that only the day before two Chinese soldiers had arrived in search of us, and not finding us there had crossed over to the Salween, happily by the Chun-tsung-la, a pass we were destined to cross five months later, located between the Sie-la and the Doker-la.

It appears that the official at A-tun-tsi had heard I was no longer at Tsu-kou, whereupon in a much perturbed state of mind he had at once sent soldiers to recall me from the Salween, whither he rightly guessed I had gone. He also wrote to the Viceroy at Yunnan-fu, who complained to the British Consul that the Chinese could not be responsible for my safety if I persisted in travelling beyond their jurisdiction. It is pleasant to be able to point out however that these representations were made entirely in good faith, and that the officials with whom I came in contact treated me with unfailing courtesy throughout.

It is but a short distance from Londre to the Mekong along the torrent which, now in full flood, poured through a fine gorge into that river. The Mekong was rising every day, its waters of a chocolate-red colour, derived from the red sandstone of the plateau, whereas the Salween was much more yellow.

A few miles higher up the river we reached Yang-tsa, and crossing by the rope bridge camped on familiar ground once more. How delightful it was to feel warm again, though the wind increased in violence throughout the evening till it blew a local gale! Then too there was unlimited fresh yak milk waiting for us, besides butter and eggs.

It is necessary of course to clamour for milk before milking time, otherwise the Tibetans, who have no use for it as a beverage, can only supply you with sour curds in lumps, like cream cheese, a great delicacy with them but not popular with me. Moreover, if they draw the milk into their own utensils, which are already filthy with the clotted cream of ages, and ever will be, seeing that they are never by any chance washed, it goes sour almost immediately, so that the operation needs to be superintended by one's own men. As a matter of fact their clotted milk is not unpalatable, if looked upon as cheese and not milk; but the uses of cheese in cookery are somewhat limited. Fresh Tibetan butter on the other hand is excellent, though combined with a good deal of hair, from being made by the simple process of kicking milk around in a yak-skin bag. Such luxuries as fresh milk and butter can be better appreciated when one remembers that in China neither butter nor milk can be obtained, because the Chinese consider it disgusting to milk cows, and one therefore has to subsist on the tinned varieties.

Two days later, that is on June 27, we reached A-tun-tsi, the journey having extended over twenty-five days, and been fairly successful in results. As in the corresponding region of the Salween valley, the weather remained fine coming up through the arid region of the Mekong, though all day long heavy masses of cloud rested on the mountains to east and west, and the usual local wind got up at mid-day without ever affecting the movement of the clouds. Never more than a few drops of rain at a time—the dregs from the cloud-fringe—fell in the valley itself at this period.

My landlord had gone away to get married during our absence, the ceremony consisting of fetching the woman and bringing her to his house, no doubt after exchanging presents with the parents. He turned up the day after my arrival, driving several donkeys laden with supplies, chiefly presents from his father-in-law. Behind him came his wife, dressed, I imagine for the first and last time in her life, in new and clean clothes, with a friend on either side holding her hands, while she coyly looked at the ground. Escorting the party came a crowd of shouting children carrying bunches of flowers, while the villagers stood around in groups to see the triumphal entry. In the evening there was a horrid orgy directly under my room, and everybody got gloriously drunk. Two days later I was called in to prescribe for the bride. On the whole, however, it was nothing like such a popular holiday as a funeral we had in the village a few months later, which was attended by the whole community.

My friend Chao, the local mandarin, called on me the day after our return and found me busily engaged amongst numerous plants spread out in all directions, which probably eased his mind. I really think he was genuinely pleased to see me safely back, and though he grumbled at my going, saying that the Lutzus were wild and bad men, I heard no more about the matter. However, he questioned me closely about my future movements, though as I did not know much about them myself, I was unable to enlighten him.

CHAPTER IX

ON THE ROAD TO BATANG—THE LAST TOWN IN CHINA

On my return from the Salween I spent nearly a month in A-tun-tsi climbing the surrounding mountains and exploring the neighbouring valleys for plants.

Every day the south-west winds blew heavy masses of cloud across, but except at night, when the valley frequently filled with mist to the accompaniment of a steady drizzle, the rain which fell was generally in the form of passing showers swooping suddenly down from the west ridge and passing as quickly. In the month of July there were only five rainy days, when it rained for six hours or more. On the mountains all round us, however, heavy clouds usually rested throughout the day, lifting for brief intervals only, sometimes with astonishing suddenness. The high range to the east was frequently covered in the evening by a long white cloud-cap, cut off sharply below, which rose and fell like the tide, exposing more or less of the mountains. Pei-ma-shan to the south-east was always obscured and drenching rain fell there every day, frequently to the roll of thunder, while bright sunshine prevailed at A-tun-tsi— a local peculiarity due to the hot dry winds sweeping directly up from the Mekong valley.

On June 30 Kin started for Wei-hsi to get me some silver and post my mail; I did not expect him back for three weeks at least. On July 5 Gan-ton also departed, as he wished to return to the bosom of his family at Tsu-kou for a week. Unfortunately he played me rather a shabby trick by staying away three weeks, so that for the rest of the month my journeys into the mountains had to be made

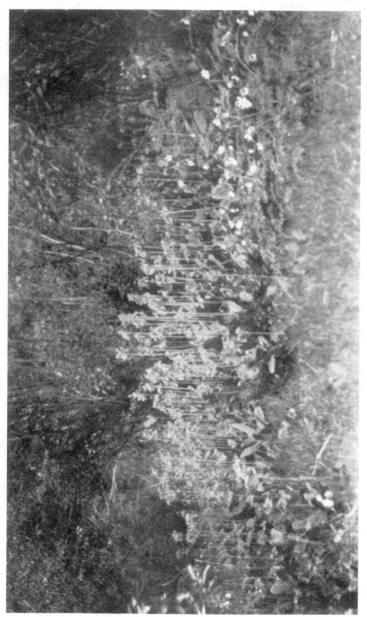

Plate XXII

Primula pseudosikkimensis growing at 11,500 feet near A-tun-tsi

alone, it being impossible to get men so late in the season. The patches of alpine meadow land which occurred here and there along the streams between 12,000 and 14,000 feet were now gay with flowers, conspicuous amongst which were *Primula pseudosikkimensis* and numerous species of *Pedicularis* with immensely elongated corolla tubes. This latter genus is indeed richly represented in these mountains, ten or twelve species at least occurring commonly, though in very different habitats—in marshes, along the edges of woods, on barren rocky slopes, and on the alpine grassland. The Himalayan region is also rich in *Pedicularis*, as it is in so many of the Western China genera, and indeed the continuity of these two floras, to which Hooker long ago drew attention, is being yearly emphasized.

To the summit of the eastern watershed was a long climb but I made it several times, finding numerous plants interesting both from a botanical and horticultural point of view. Immediately above A-tun-tsi came a shrub belt consisting almost entirely of scrub oak on the slopes exposed to sun and wind, but on the moist shady slopes exhibiting a rich assortment of *Cotoneaster, Salix, Rhododendron, Populus, Hippophaë, Philadelphus, Deutzia*, with many beautiful roses such as *R. sericea*, and a considerable herbaceous undergrowth. This undergrowth included, besides the plants previously mentioned, a yellow violet (*Viola Delavayi*), the *Podophyllum* already referred to, whose big pear-shaped fruits (which take four months to ripen) were beginning to turn red, and a very sweet-scented *Pyrola* (*P. atropurpurea*).

Above this was a narrow forest belt of spruce, but including also larch and birch trees on north-facing slopes, where the snow melted more gradually; and above this again came the shrub belt of the alpine region, composed almost entirely of scrub rhododendron and a cream-flowered potentilla (*P. fruticosa*). This ended abruptly on the screes of the exposed slopes, but elsewhere it dwindled gradually into alpine grass-land, rich with saxifrages, gentians and other flowers, extending to about 18,000 feet. Up here, at 17,000 feet, springing from amongst huge blocks of grey stone, I found the glorious Cambridge blue poppywort (*Meconopsis speciosa*), one of the most beautiful flowers in existence, several Primulas, and a number of highly adapted plants inhabiting the screes and the icy puddles of water which trickled from the melting snow.

Two varieties of *Lilium lophophorum* found on the alpine grass-land at 14,000–16,000 feet were peculiar in that the corolla was pendent, with the tips of the petals cohering, but the fruit erect. These flowers open in the rainiest month, and the reason for their being pendent may well be to preserve the honey and pollen, to which insects have access through the slits between the corolla lobes. The fruit, on the other hand, ripens during the dry autumn, and stands erect on a resilient pedicel, the winged seeds being shaken out in the gales and carried far and wide. This arrangement of pendent or horizontal flowers is very conspicuous in the summer flora, especially on the very wet Salween-Mekong watershed (e.g. *Primula Souliei, Meconopsis pseudointegrifolia*, etc.) ; but the autumn flowers on the Yang-tze-Mekong watershed (saxifrages, gentians and so on) stand erect, so that I regard it as an adaptation to meet climatic conditions.

From the time the snow disappears towards the end of May till the grass withers and winter sets in about the end of September, the herdsmen camp in the high mountain valleys, fattening their flocks of sheep and yak on the rich alpine grass-land, from 14,000 to 16,000 feet above sea-level. Once or twice Gan-ton and I had been caught in heavy rain storms when high up in the alpine pastures, and had resorted to these herdsmen's tents for shelter and refreshment.

A small ridge-tent of brown hemp cloth, the sides pegged down and weighted with stones, one end built against a rock or stuffed up with branches, the other open to the winds—this is the home of the Tibetan herdsmen for four months of the year, while his food consists of *tsamba*, tea, butter, and sour milk. There is just enough room for three or four to squat cross-legged round the fire in the middle, which fills the tent with pungent smoke. The remaining space is occupied by the leather bags of *tsamba*, the wooden cylinders for making tea, and the wooden milk-pails, so dirty with clotted curds that, as I have already said, milk drawn into them rapidly turns

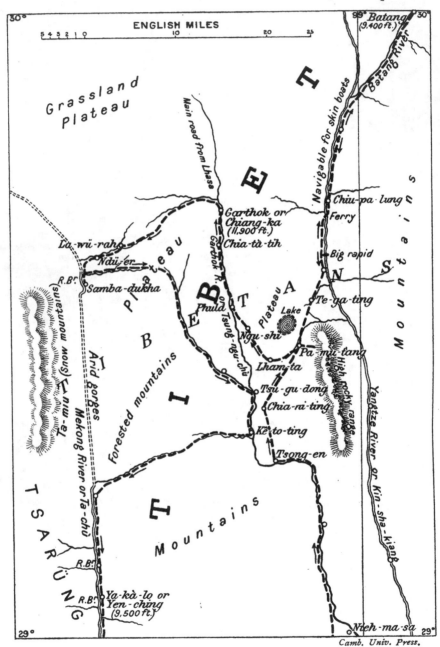

Map V

ENGLISH MILES

Sketch map to illustrate journeys to and from Batang.

[*Chapters IX to XI.*]

[*R.B^r. = Rope bridge.*]

sour. The men dress in the skins of animals and huddle by day round the fire, sleeping at night on beds of pine-branches. An altar is always rigged up at the far end of the tent, and here a single butter lamp splutters, faintly illuminating small offerings of *tsamba* or barley grains, and clay ikons of the crudest form, daubed with pats of butter.

Kin one morning watched a herdsman rush out from a tent with his long gun, kneel, and fire at a dark object which was moving coolly up the mountain slope with a lamb in its mouth. It was a leopard. These marauders, which stalk the mountains in broad daylight, are a constant source of alarm to the herdsmen, though they never dare attack any but isolated animals. At night they descend to the lower valleys, several having been reported in the neighbourhood of A-tun-tsi while I was there, and in the winter they come right down into the village, though I never saw one myself. Deer, however, and precipice sheep I saw on several occasions, and sometimes when camping in the forest I would awake in the dead of night to hear Ah-poh barking furiously at the entrance to my tent, as some denizen of the mountains prowled by.

Meanwhile Kin had returned after an absence of exactly three weeks. He brought me silver and welcome letters, and reported heavy rains in the Mekong valley below Tsu-kou, several of the rope bridges being under water and impassable.

On July 21, a great Mohammedan festival known as the *ho-pa-hwei* was held in A-tun-tsi in honour of a certain Ming Emperor, called *Pei-wang* or the White King, who came from Tali-fu.

Outside many of the houses, torches ten or twelve feet high had been built by tying bundles of pine-sticks in tiers round a central pole, the entire structure being decorated with flowers, branches of green leaves, and paper flags, making a gay show. As soon as it was dark crackers were fired as a signal for the revels to begin, and immediately afterwards the big torches were lighted at the top; and looking down the street one saw by the light of these beacons which smoked and crackled on either side, the black figures of people dancing.

Everyone was out of doors. Processions of boys formed up and ran round the village, and so along the hill paths above the cultivated slopes, waving fire-brands and whirling round glowing sticks snatched from the torches. The principal Mohammedan merchants had decorated and lighted up the family altars, and engaged musicians to beat drums and cymbals to exorcise all the devils which had gathered during the year, and the din went on all night. Large grotesquely-swollen lanterns swung to and fro in the evening breeze, feasting was carried on till a late hour, and everybody got very drunk in honour of the White King. Altogether it was a most successful carouse.

Next day we started for the pass on the road to Pang-tsi-la, as I wanted to climb the lower slopes of Pei-ma-shan, the big snow-mountain on this watershed. The difference of climate should, I thought, make a great difference to the flora, but here I was mistaken, as it subsequently turned out. The same plants flowered on the Pei-ma-shan range one or two months earlier than on the mountains in the immediate neighbourhood of A-tun-tsi, and though in the course of time the cumulative effects of this early flowering may differentiate the floras of the two districts—distant only six miles as the crow flies—by bringing them into contact with different sets of insect visitors, I did not observe that this had as yet happened.

It was beautiful weather when we left, but it was not long before we got fairly into the rainy region, and drenching showers fell throughout the day. Riding along in front of the slowly-moving line of porters, thoroughly wet and chilly, I watched for my friends the animals, and noted the changes in the vegetation as we ascended. I saw a fine badger, in spite of the daylight, dash out from the forest, and later a chipmunk running about in a tree, but there was little else and birds were as usual scarce.

There were many beautiful shrubs in flower, however —*Viburnums*, pink and white roses and hydrangeas, and various species of *Prunus*, *Ribes*, and *Rubus*, with great splashes of white clematis trailing everywhere, and by the torrents a rich undergrowth of ferns, including the maiden-hair. Tall purple-flowered meadow-rues and brilliant blue monkshoods grew by the wayside, and on the drier slopes were stately crimson spikes of *Epilobium*, with an occasional white-flowered specimen.

It was dusk before we found a convenient camping-ground by a small stream. Everything was sodden, we were still some distance from the pass, and there was no prospect of getting a good fire or having anything hot to drink for some time. However, we put up the tents, tethered the ponies, and did what we could with the damp firewood.

Next morning, leaving the men in camp, I climbed to the top of the pass, which crosses the Mekong-Yang-tze watershed at 15,800 feet, and from there ascended the ridge, which runs south-west towards Pei-ma-shan. A great plateau country covered with dwarf rhododendron spread away to the south towards the snow mountains, which were entirely obscured, and in the shelter of the screes vegetation extended to about 18,000 feet, where various primulas, saxifrages, and *Meconopsis speciosa* grew scattered amongst the rock fragments.

On the other flank of the pass, however, bare screes faced the south, the belt of dwarf rhododendron ending a few hundred feet above the pass, while above towered the curiously fretted limestone buttresses of the divide. Throughout the day a keen wind blew down from the plateau bringing showers and mist, sometimes in the form of a drizzle, sometimes in storms of chilly rain.

The valley facing A-tun-tsi which we had ascended owes its elevated tree-limit (15,000 feet) and the presence of such trees as the larch, to the local down-valley rain-bearing wind which drenches the valley almost every day throughout the summer, while the A-tun-tsi valley itself is swept by the direct up-valley wind from the Mekong. In the mountains here were many signs of previous glaciation —stair-way structure in the valleys, piles of angular scree material having the appearance of moraines at the valley heads, and a curious plateau structure at the foot of the peaks dominating the valleys, across which the collecting streams wandered independently, sometimes forming small lakes, before joining to dash down the stair-way. But the well-defined rock basins to be described later were wanting.

South of the pass the range was capped by granite and metamorphic rocks, in strange contrast to the limestone met with all along the summit of the range northwards.

The weather showed no signs of improvement on the following day, and I went up the nearest valley to the north in order to examine the scree flora of the south-facing slopes.

These screes support a few highly specialised plants, which exhibit morphological peculiarities deserving of mention. Their root-systems, which are of immense length, are often strengthened with T-shaped bracing pieces, while their stems have scarcely-developed internodes. Their leaves too are either succulent, or hairy, or red ; and occasionally peculiarities are noticeable in the arrangement of the inflorescence. I also found several plants assuming the cushion form, a habit commonly met with in deserts such as those of Algeria, including a species of *Potentilla*, a *Lychnis*, *Diapensia Bulleyana*, and other species of Caryophyllaceae. But these dreary screes were practically devoid of plants suitable for horticulture.

Above the valley head dismal wastes of rock rose steeply to the limestone towers crowning the ridge, but the valley floor was also strewn with big blocks of a quartz breccia, cliffs of which were visible lower down. Late in the afternoon I returned to camp, and a glorious evening set in ; so cold was it in the night, however, that even inside my tent the iron supporting-pole was wet with dew.

I had injured my foot on a rock some days previously, and as it refused to heal but got rather worse, and was now very painful, I decided to break camp and give it a day's rest at A-tun-tsi.

Starting early next morning and leaving the men to pack up and follow, I rode into A-tun-tsi at one o'clock, to find the village bathed in brilliant sunshine ; yet in the afternoon the sky grew black over Pei-ma-shan again, and we heard the thunder rolling and rumbling over that storm-riven mountain. At sunset the peaks, clearly outlined against the blue-black sky, presented an extraordinary appearance, as though recently swept by a terrific snow-storm, for they glowed with a pale reddish-gold tint, which in contrast to the darkling sky and surrounding mountains, now in deep shadow, looked like snow. Then the stars came out in their myriads, and distant lightning could be seen flickering behind Pei-ma-shan.

July 27th, two days later, was a perfect day. From the time the sunlight flooded the valley at ten o'clock till it sank down behind the lonely monastery at four there was not a cloud in the sky.

In the evening Gan-ton, whom I had given up as a deserter, arrived unexpectedly from Tsu-kou, bringing me a message from the French priest; and as I read it I realised that an entire change of programme would immediately be necessary. The message read as follows :—

> "The English are in Lhasa, the Chinese soldiers have capitulated....A British officer has gone in from Y'a-k'a-lo on a secret mission. The Chinese are furious and swear to exterminate every Englishman. I fear you will be killed before the end. You must leave A-tun-tsi at once."

The situation demanded action, not thinking over. But what? The story seemed on the face of it improbable, yet it was not a message to be entirely disregarded. To sit still and wait for something to happen was against all my principles, and I decided to follow this advice and leave A-tun-tsi immediately.

But there was no object in going south down the Mekong, for if true, the story would be known throughout southern Yunnan long before I could reach T'eng-yueh, while on the other hand if it were not true, I ran the risk of ruining a year's work for nothing. I made up my mind therefore to go still further into the country, and within ten minutes of receiving the message I told the men that we were starting for Batang the first thing in the morning.

The advantages of going north were obvious. In the first place, Batang being on the main road to Lhasa, I should there hear the truth of this wild rumour from Tsu-kou, if indeed there was any truth in it. Again, there were several European missionaries at Batang, and it would be best in the event of trouble to band ourselves together, while as regards routes Batang perhaps offered a greater choice than did A-tun-tsi. Finally, by going north I should be keeping within the chosen area for botanical work, and might reasonably hope to find many of the plants I had already observed

on the Mekong-Yang-tze watershed; so that in the event of
my return to A-tun-tsi being delayed I might still secure seed
of many species already noted.

Everything was packed that evening, and Kin went down
to the yamen for *ula*, telling the official that I was going
into Ssü-chuan. No objection was raised, and I dare say
the official was rather glad to get rid of me for a bit, though
he showed no sign of having heard anything about Tibet.
I told the men nothing, merely saying that I was going to
Batang, and would be back in two or three weeks. Gan-ton
was to accompany me, and travelling very light I hoped to
do the eight stages by the direct road in five days, Kin
following more leisurely with the camp equipment as soon
as he could secure the necessary animals. The bulk of my
luggage was left in charge of Sung, who was deputed to look
after it till my return.

I had received the message at five o'clock in the afternoon and by eight o'clock next morning we were on the
road to Batang, At such short notice the yamen had only
been able to provide me with one pony, and the light loads
were carried by three porters, a solitary soldier also accom-
panying us as escort. I brought with me only a little
bedding, and one box, containing such necessaries as my
passport and cards, a map, aneroid barometer and compass,
a medicine chest and flask of brandy, a few tins of meat
and other stores, silver, and photographic outfit. Finally
I slipped a heavy Colt automatic pistol with spare magazine
and cartridges into my pocket.

Crossing the spur just above A-tun-tsi we descended
the long winding valley through beautiful woods to the
village of Adon, situated at the point where a mountain
torrent, after flowing southwards for some distance, suddenly
turns westwards and breaks through a deep gorge in the
barrier range to reach the Mekong below. They were
reaping the wheat and barley here already, and the pears
and apricots were fast ripening, for Adon is between two
and three thousand feet below A-tun-tsi.

At the house of the chief we stopped to change the *ula*
and have lunch, ponies being supplied in place of the porters.
The old chief had been beheaded by the Chinese after the
1905 troubles, and his son, a lad of fourteen, was now at

school in A-tun-tsi, though the big house was still kept up. The chief of the next village was also rather juvenile in years, if not in experience, for though only sixteen and still a schoolboy, he already boasted two wives.

The large Tibetan farm-houses, or manors, of the mountain villages are always widely separated, and towards the limits of habitation in the high plateau valleys are completely isolated. The thick walls are built of mud, stamped hard and whitewashed on the outside, but the supporting-pillars for the roof, the floor beams, window-frames, and partition walls are of wood. It is curious that though glass is unknown, excellent window-frames should be built.

On entering from the outer courtyard one finds oneself in a large gloomy stable occupying the entire ground floor; to the wooden pillars cows and ponies are tied, and what light there is comes from a square hole in the roof, through which we climb up a notched tree-trunk, in place of the more familiar ladder.

Above is a verandah, and from it the spacious rooms open—the big kitchen and living-room of the family, guest-rooms, prayer-closets, store-rooms, and so on ; but the kitchen, as the most important room of the establishment, is by far the most interesting.

At one end is a big open hearth on which a wood fire is blazing, the smoke finding its way out by a square hole in the roof or through one or two small windows. Every beam and rafter is black with soot, but as the Tibetans always sit on the floor, never on benches, they are not inconvenienced by the smoke curling up above them. Rows of copper kettles and brass jugs stand on the shelves, several bamboo spears are leant against the central wooden pillar, an iron pot bubbles over the fire, and tall brass-bound wooden cylinders in which tea is made as already described, stand on the floor. Roof, walls, and floor are hideously filthy with dust and soot, but the brass-work is kept surprisingly clean. Behind the fire the wall is often crudely decorated with Buddhist designs in whitewash, and here in a dingy alcove is arranged the family altar, with small metal vessels full of grain, and butter lamps burning dimly. There is no furniture unless one or two rough window-seats may be described as such, and no other decoration, yet, in the darkness which hides the dirt, the size of the room and its severe simplicity give an imposing effect. It is warm in winter and cool in summer, though the myriads of flies make it extremely disagreeable. The roof, which is flat and paved with a hard mud floor, is reached through another smaller aperture placed to one side, and the roof-garden, so to speak, is used for threshing corn. I use the term roof-garden advisedly, because I have sometimes seen boxes of flowers lining the low parapet which surrounds it, though this is not common.

An open shed in which the corn is harvested extends along one side, and a dummy chimney into which are thrust numerous bamboos bearing prayer-flags and strips of white paper or silk decorates one corner.

These big houses as a rule harbour more than one family. The architecture is what might literally be described as severely perpendicular, and they are built primarily I imagine to withstand and keep out the howling winds which make life on the plateau so rigorous. Many of them must be extremely old—I never saw one in course of construction, and the timbers appear so well seasoned that they should be eminently capable of standing the ravages of the weather.

In the evening, just as the rain began, we took shelter in the last house up the valley, and I found myself installed in the open shed on the roof, where I was surrounded by sheaves of corn. A big wooden comb stood on the ground, and women had been busy pulling the wheat through this in order to separate the heads from the stalks preparatory to threshing.

Starting at six o'clock next morning we climbed all day, reaching the head of the pass known as the Tsa-lei-la (15,800 feet) towards evening; there was a fine view of the Salween divide and K'a-gur-pu to the south, but to the north the sky over Batang was a lowering blue-black.

Crossing this pass I was for the first time struck by the great difference of vegetation exhibited by valleys facing north and south, as well as by the even more striking differences, both botanical and geological, between the Mekong-Yang-tze and the Mekong-Salween watersheds.

Ascending the south slope we found at the limit of trees spruce and *Juniperus*, the latter not like our English juniper bushes, but big trees. These ceased at an altitude of little over 14,000 feet, and alpine grass-land, with narrow strips of dwarf juniper under the protection of the rocks, extended to the summit. On the north-facing slopes, however, spruce and larch were the limiting species, extending to within 600 feet of the summit, to be succeeded by a thick belt of dwarf rhododendron which scarcely died out before the summit was reached, the belt of alpine grass-land being correspondingly reduced.

This marked peculiarity of south-facing valleys is due, firstly to the more rapid melting of the snow in spring, whereby the only available supply of water is removed at the very outset of the vegetative season, and the un-protected young shoots suddenly exposed by the rapid stripping off of the warm snow blanket; secondly to the havoc wrought by the dry, rainless local winds to which the south-facing valleys are liable throughout the summer.

Looking eastwards from the pass, there came into view an extraordinary wall of limestone towers and buttresses crowning the Mekong-Yang-tze divide. These splintered cliffs and bare screes testified to the weathering work performed by dry denuding agents—a wide range of tem-perature with rapid alternations, the effects of wind, the splitting action of frost and so on—and were in strong contrast to the graceful pyramids carved out by water erosion on the far more rainy Mekong-Salween water-shed.

This chain of limestone towers can be traced from Pei-ma-shan in latitude 28° 15′ to latitude 29° and between these two points we crossed the watershed by four different passes. Indications of previous glaciation were everywhere visible, and there can be little doubt that this ridge has had its rainfall enormously curtailed here by the very consider-able elevation of the Mekong-Salween watershed further west, which thus, as previously stated, intercepts the bulk of the monsoon rains from the south-west.

The descent to the Tibetan village of Tsa-lei occupied nearly three hours, and it was dark when we arrived after fourteen hours in the saddle.

Tsa-lei, situated at an altitude of nearly 13,000 feet, is built on the edge of a small mountain-flat which has evidently been formed by the silting up of a lake. Numerous torrents converging on to this grassy pocket have thrust out alluvial fans on which grow dense thickets of *Hippophaë rhamnoides*, a very characteristic formation, for this tree grows along all the more sluggish streams of S.E. Tibet, often in dense thickets.

It felt very cold in the night at this altitude, for the huts were of pine logs with shingle roofs, such as the Lutzu build; indeed all up the Mekong valley, even into Tibet itself, this style of architecture is adopted by the poorer people.

Though we were up at five next morning we did not get off till nine, for all the animals were up in the mountains for the summer, and we had the utmost difficulty in securing two ponies and a yak. These ponderous yaks trudge along at a terribly slow pace, but they have their advantages in the summer at least, when swollen torrents have to be crossed. More than once I thought my pony would be swept away by the rush of water, when my feet were only just awash, but the stumpy-legged yak, though he dipped his loads every time, swept through the water like a snow-plough.

The Garthok river, a tumultuous red flood flowing through an arid gorge, was reached just as a terrific rainstorm began, and though we continued our march till dark, the morning's delay and a good deal of time wasted in changing the *ula* at mid-day considerably curtailed the stage. Crossing the river by a most crazy wooden bridge we stopped for the night at a small village on the opposite bank, and were no sooner inside the nearest hut than it again began to pour with rain, continuing without intermission all night.

After supper as I lay on my mattress in the tiny room allotted to me writing up my journal by the light of several pine-wood chips blazing on a stone, in stalked three Tibetans, all of them over six feet high. Their coarse gowns were tied up above their knees, the right shoulder thrust jauntily out exposing the deep muscular chest, and they were bootless. One of them carried a fiddle, consisting of a piece of snake-skin stretched over a bamboo tube with strings of yak hair, upon which he scraped vigorously with a yak-hair bow.

There was little enough room, but my visitors soon lined up, stuck out their tongues at me in greeting, and began to dance, to and fro, up and down, twirling round, swaying rhythmically to the squeaky notes of the violin (there were only about two notes on which to ring the changes), and singing in high-pitched raucous voices. Presently three women joined in, all tricked out in their best skirts and newest boots, with cloaks flung negligently over their shoulders. Thus they went through many of their national songs and dances, and in justice to my sex I must say the men danced with more skill and grace than did the women, though of course it is easier to dance heel and toe, bare-footed like the men, than in the clumsy boots and skirts worn by the women.

I can still picture the scene in that dim little smoke-blackened room, the rain lashing down outside, and the roar of the river just below us, while I lay back on my bed enjoying it hugely, all cares forgotten. Those great giants of men looked strangely weird in the flickering light of the blazing torches which flared up and burnt down alternately; the wail of the fiddle rose and fell, the voices blended, and broke, and ceased, and still they danced on, up and down, to and fro. They danced for two hours in all, and in return for the little present I gave them would willingly have gone on till midnight had I not told Gan-ton I wanted to go to sleep.

The Chinese, so far as I know, have no country dances like these, indeed they do not dance at all, and would consider any such mingling of the sexes on terms of equality highly improper. Even in the theatres women are not allowed to act with men, their place being taken by men dressed for the part.

At breakfast next morning I was twice startled to hear the rattle of falling rocks, and looking across the ravine I saw a cascade of bouncing boulders pouring down the steep cliffs into the boiling river, just where the road passed underneath. It was still drizzling when we started at seven o'clock, and we hurried the animals one at a time past the danger zone, where the path had been ploughed up by the falling rocks. Following up the arid gorge of the Garthok river till mid-day we finally struck up towards the plateau, passing through a country of red sandstone, much dissected by torrents, which flowed in deep straight-sided gullies. At the village where we halted for lunch was a small lamasery, much decayed.

The hill sides here were very barren, probably visited by rare but furious rainstorms, which, leaving little time for the water to sink into the soft sandstone, tear deep rents in it, and flowing into the Garthok river below, colour it a light chocolate red. This sandstone forms a very important feature of the scenery in S.E. Tibet, and I traced it right across the gently undulating plateau country which I traversed.

The lower mountain slopes were strewn with boulders of harder rock, evidently carved out *in situ*. Sills of this rock, a closer-grained sandstone, or perhaps an iron-stained limestone, stood out boldly in many places, forming conspicuous ledges, and being well jointed, they often exhibited an incipient hewing into separate blocks. Cliff sections sometimes showed large boulders embedded in the soil, and all doubt as to their origin was removed when I saw blocks of stone only recently detached from a sill lying about in all stages of exfoliation from cubes to spherical boulders. The obliteration of Tibetan inscriptions cut on way-side rocks, many of which were certainly recent, since the colours in some cases had not had time to be effaced, testified to the softness of the sandstone. Higher up in the hills trees appeared—willows, *Thuja*, *Hippophaë*, and so on, with patches of grass-land and fir forests.

Later in the afternoon it began to rain with amazing violence, the slopes streamed with water, and the ponies slipping dangerously on the slimy red surface, we had to go dead slow. At six o'clock we reached the miserable village of Chia-ni-ting utterly drenched, and took refuge in the largest house we could find. It was so cold that I was glad to have a fire, but it went out during the night, and being unable to sleep I got up at half-past three and roused Gan-ton, who set about preparing breakfast.

Plate XXIII

Sunset over the Yang-tze; looking
north to Batang from the edge
of the plateau

Presently our soldier, who had slept like a Trojan, came in rubbing his eyes, thoroughly disgusted at the early start contemplated, but like the good fellow he was, he made himself useful and we started once more at five o'clock.

The misfortunes on the third day had lost us a lot of time and it was now impossible to reach Batang in five days, but a big effort might bring us there on the sixth day, which was good travelling, for it is eight mule stages from A-tun-tsi by this road.

We found it bitterly cold as we scampered over the plateau on that raw misty morning of July 31, 14,000 feet above sea-level. To the west, the low rolling hills stretched away as far as the eye could reach, brilliant green turf, red sandstone cropping out here and there, and patches of vivid blue where some Boraginaceous flower grew in dense masses. Hundreds of little pikahares darted into their burrows as we rode along, and, in spite of the severe cold, the day promised to be fine. Down in the valley below were meadows of flowers—blue salvias, monkshoods, and borages, crimson and yellow Scrophulariaceae, and various other plants, but I saw nothing exceptional here.

We changed animals no less than five times in the course of the day, wasting a considerable amount of time, but nevertheless we rode on till nine o'clock at night, by which time we had reached the Kin-sha, or Yang-tze. Never shall I forget the first view of that noble river as we climbed the last spur and looked northwards over the trees towards Batang. The sun was down, and over the purple mountains great puffs of radiant cloud rested, scattering the dying light; for miles we could follow every twist of the valley, marked by a ribbon of flashing silver, which had still 3000 miles to flow before it reached the ocean. The descent occupied two hours, and the abrupt change from the bitter cold of the rain-swept plateau which we had experienced in the morning to the hot breath of the wind blowing up this gutter, was one of the many extraordinary things in this extraordinary country.

On the way down to the Yang-tze I came across the beautiful scarlet *Androsace Bulleyana* with dense heads of flowers, growing in the very driest places, and the rocks of the main valley were covered with the lovely asphodel,

Eremurus chinensis, sending up tall spikes of densely-packed white flowers from a rosette of bayonet-like leaves.

Gan-ton and I were in excellent spirits as we rode along the narrow winding path by the light of the young moon, and in spite of nearly fourteen hours in the saddle he began to talk of Lhasa, five weeks' journey west of us, and asked whether I would like to go there! Down below in the darkness the river could be heard booming over the rocks, and the raging wind from the south grew stronger and stronger, though it did little to cool our parched skins. At last after the moon had set, we saw lights ahead, and going now very slowly in the darkness we came upon a group of houses and quickly gained admittance. Everybody was asleep round the dying fire in the big kitchen, but Gan-ton turned them out, and they gave me a room and some hot milk, blinking and rubbing their eyes as the pine torches fizzled and blazed up. I turned in soon afterwards, but finding the sudden heat rather trying, I had no difficulty in getting up at half-past three next morning, when the air outside was deliciously cool.

We were off at five, at which hour the deep valley was slowly filling with a sickly grey light; the wind, which had gone on increasing in violence till midnight had then ceased abruptly, and everything was very still. The size of the Yang-tze now impressed me greatly; the Salween and Mekong looked puny compared with the storm of water which came smashing down over some immense boulders where a heavy landslip had choked its bed. In winter the river would certainly shrink a great deal, but even so its source must be many hundreds of miles from here, and it is undoubtedly as regards length the first river in Asia. Arid as the valley was, the herbaceous vegetation seemed richer than in the Salween and Mekong gorges. Species of *Androsace, Eremurus, Allium* and other Liliaceae, two species of *Clematis*, one a small erect shrub with white flowers (*C. Delavayi*), the other a twiner, but often growing erect or procumbent on the sand dunes where no support was available (*C. splendens*), were common, in addition to other plants peculiar to these semi-desert valleys.

There being no rope bridges over the Yang-tze, we crossed in a big scow propelled by clumsy oars and a huge

Plate XXIV

Tibetan women of S.E. Tibet, showing the curious manner of doing the hair

A Tibetan girl of A-tun-tsi

stern-sweep, being washed down a quarter of a mile during the passage. Then came a long hot ride as the sun rose higher and the valley became more and more shut in. In the afternoon we left the river and crossing a high spur looked right down on to the little mountain plain of a few hundred acres, where Batang stands. There were numerous caravans coming in and the road was blocked with mules, but Gan-ton and I, dashing ahead, rode into the city at five o'clock.

Numbers of Tibetan girls and priests were lounging about, gossiping and washing their clothes in the stream, as we crossed the little stone bridge and rode up the steep cobbled street to the mission house ; and every one stared at us in surprise. We had covered the 180 miles of mountain road in six days.

CHAPTER X

ACROSS THE CHINA-TIBET FRONTIER

THE first person I sought in Batang was Mr Edgar
of the China Inland Mission, to whom I confided the
disquieting rumour which had reached me at A-tun-tsi.
He had heard nothing of it, however, and everything was
going on normally. Nevertheless after tea we went round
to see the French priests, who should know if anybody, since
the story had originally come from one of their number;
but they too denied all knowledge of the matter, and it
was obvious that there had been a mistake. We then
telegraphed to the English Consul at Chengtu, but no
answer had come when I left four days later, though
subsequently the Consul wired back that there was no
truth in the rumour.

Meanwhile I was the guest of Mr and Mrs Edgar, and
enjoyed a well-earned rest. We talked geography till far
into the night and it was now that I learned who were
the mysterious Europeans at Men-kong before me. Several
months later we heard how Captain Bailey had successfully crossed to India, and at the same time gathered the
foundations of the strange story from Y'a-k'a-lo, which
may be interpreted as follows.

It appeared that Captain Bailey had with him a Chia-
rung tribesman whose knowledge of Tibetan was of the
scantiest. This man had been sent back to China from
the borders of the Mishmi country, and while talking
volubly of the prowess of his master had so mixed up
his tenses that the story of how Captain Bailey had been
to Lhasa with the British Expedition acquired present
significance. This of course is only a surmise, but it
accounts for the facts.

Moreover this Chia-rung tribesman seems to have been of a highly original turn of mind, for after dashing through to Men-kong, exclaiming wildly all the way that he had been ordered to annex the country for Great Britain, he so scared the officials in Y'a-k'a-lo with his strange tales that they began to suspect he had murdered Captain Bailey; and no sooner did he get back to Batang than he was clapped into prison, where I believe he lingered for a month.

Batang is situated at an altitude of 9400 feet, and the little plain being closely invested by mountains grows very hot in summer, though it is not cold in winter. A gentle breeze frequently sweeps down from the high ranges to the north-east and fans the parched earth, but on occasions the valley is swept by fierce gusts blowing up the Yang-tze. The population now comprises between 400 and 500 families, and since the rebellion of 1905 from being almost exclusively Tibetan, with all the power in the hands of the lamas, it has become very largely Chinese, and the power of the lamas is temporarily broken. On the other hand the majority of the Chinese, merchants and soldiers, have married Tibetan wives and adopted at least some of the manners and customs of the country if not the dress.

Crops of maize, wheat, and barley are grown, besides buckwheat in the autumn, but the area under cultivation is very small. Many of the houses are built of stone, and there is an air of prosperity about the place, with its streets of shops and hawkers, in spite of the gaunt skeleton walls of the once huge monastery, now utterly destroyed. Since the rebellion, the majority of the lamas have been killed or scattered, and the ragged-looking mendicants who now hang about the streets or loaf round the tiny lamasery which the remnant were allowed to rebuild, are no credit to the profession.

I have already referred to Chao Er-feng, Warden of the Marches, and subsequently Viceroy of Ssü-chuan, who was entrusted with the stamping out of the Tibetan revolt of 1905; and however much one may denounce his methods, he met with considerable success. Peace and security now reign in Batang (or did before the present Revolution) instead of lawlessness, robbery, and murder. Unfortunately

Chao was not content with securing safety along the main road across Ssü-chuan as far as the Tibetan frontier, but must needlessly push on to Lhasa, an unjustifiable procedure, since he had no money. To meet his requirements he seized the funds of the railway syndicate to pay for this vainglorious adventure, and it was owing to this that the dissatisfied people of Ssü-chuan rose in the capital shortly after I left Batang.

Meanwhile Chao had accomplished much, having established the telegraph line and postal communication between Batang and Lhasa, though it is easy now to see the tremendous nature of the task undertaken by the Warden, and the folly of embarking on such an ambitious adventure with so much discontent behind him. Far Western China and Tibet are so completely severed from China proper in community of interests, by geographical and physical barriers, by race and creed, that an independent western China seems inevitable. As an *active* colonising power, China has proved herself a failure, owing to her haughty attitude, her lack of sympathy with the natives with whom she comes in contact, and her unique methods of administration. As an indirect colonising force, however, by means of passive absorption, the Chinese, owing to their extraordinary adaptibility and virility, are unsurpassed.

Chao was murdered soon after he became Viceroy of Ssü-chuan; the grim old Warden of the Marches who had not scrupled to slay and hammer with wooden clubs those who thwarted or opposed him, who had intrigued with the Tibetans to set up independent government in Tibet, who had harried the lamas and razed their monasteries to the ground, was the victim of conflicting interests. He paid the penalty of the autonomy enjoyed by the great western province of Ssü-chuan throughout the Tibetan campaign, being ultimately murdered by his own soldiers.

While in Batang I was privileged to see one of the minor Tibetan princes who a couple of months previously had received 1500 blows by order of the Viceroy for misgovernment. The wounds were now almost healed, though the man showed us a round hole the size of a crown piece in his thigh, and was scarcely able to walk. As a rule 1500 blows such as the soldiers give with heavy wooden clubs is sufficient to flay a man to ribbons, and, hammered till he resembles a senseless jelly, he is dragged out to die; but this sturdy man had lived, and Mr Edgar had visited him day after day to pour balm into his wounds, though no doubt he was a scamp.

In appearance he was a well set-up good looking young fellow, and his small room was surrounded with the most gorgeous knick-knacks of their kind—silver prayer-wheels and *tsamba*-bowls, a silver-mounted snuff-box made from the claw of a tiger, brass images and kettles, and alabaster cups. Round his neck was an amber and coral rosary; on the walls were quaint pictures emblematic of Buddhism and horoscopes, and the floor was carpeted with rich rugs.

On the third day of my stay at Batang, Kin arrived, so on August 6 we started back for A-tun-tsi, not however by the most direct route, for I had applied to the Prefect for permission to go westwards by the main road into Tibet, a request which he reluctantly granted on condition that I would sign a paper relieving him from all responsibility.

The following statement was therefore drawn up in English and Chinese, two copies being made, one of which I retained myself.

"I, F— K— W—, wishing to go to Garthok in Tibet for the purpose of collecting plants, go entirely on my own responsibility, nor do I ask for any help from His Excellency W—, Prefect of Batang."

Probably the Prefect realised as well as I did that so far as our respective Governments were concerned, such a document was not worth the paper it was written on, and I was astonished at his accepting this guarantee; on the other hand his obligation ended there, and it was entirely an act of grace on his part to provide me with an *ula* passport and a mounted trooper as escort. It appeared later that the Prefect had made this provision never dreaming that I would agree to sign such a document, for no sooner had I departed than he fled to Mr Edgar in a most pitiable state of nervousness, begging him to call me back; but it was then too late.

For the first two days on our return journey to A-tun-tsi we retraced our footsteps, but on reaching the Yang-tze, we abandoned the long hot ride down the valley, and embarked in a coracle for a voyage down the river. This coracle was made from three ox-hides sewn together and caulked. The framework consisted of four pieces of wood in the form of a trapezium, from which the hard semi-globular skin was suspended like a bag, expanded below by means of a few lianes stretched across like ribs. Its extreme length was barely six feet, its breadth three feet, and its depth two feet, so that with five men and the luggage on board there was not much room to spare, and the gunwale was only a foot above the water. It looked the flimsiest thing in the world to launch on that great river, but its strength and seaworthiness were beyond question. The Tibetans navigate many furious rivers in these craft; but when they are very drunk it is best not to sail with them, for then they fear nothing.

A single Tibetan formed the crew of our coracle, and leaning over the narrow forward end he from time to time dug deeply into the water with his paddle, though only for the purpose of keeping us well out in the current. Beyond that he made no attempt to steer, and we drifted lazily down the river, sometimes broadside on, sometimes spinning slowly round, according to the caprice of conflicting currents. When the man put us ashore some fifteen miles lower down—it had taken little more than two hours—he drew the coracle ashore and picking it up, started off homewards with it over his head and shoulders, like a gigantic hat.

It may be worth while to record here a few differences I remarked between the Yang-tze at Batang, and its sister rivers the Mekong and Salween. A few miles lower down, it is true, the Yang-tze is less unlike them, for it also flows through a series of terrific gorges; but even there it is sufficiently dissimilar in important respects to make the following comparison of interest, and this change of character itself at once distinguishes the Yang-tze from the Mekong and the Salween.

(i) The Yang-tze is much broader than either of the other rivers, probably averaging nearly twice the breadth of the Mekong.

(ii) The mountains rising immediately above the river

Plate XXV

Yak-skin coracle on the Yang-tze, near Batang

Lake occupying a rock-basin at 16,000 feet on the Mekong-Yang-tze Divide

are neither so high nor so steep as in the case of the other two[1].

(iii) It has the slowest current of the three, though it brings down much more water than either of the others.

(iv) It flows at a considerably higher level. This is a point to which we shall have occasion to refer again.

(v) It is interrupted by fewer rapids. From the Batang river to below the ferry, a distance of quite twenty miles, there is not a single rapid which would impede navigation by canoe.

(vi) A continuous shelf of alluvial detritus and blown sand forms a low platform or bank between the river and the mountains, so that on either side the path rarely need ascend to any height above the river. Small sand-dunes are not uncommon. Houses are not, as on the other rivers, confined to the mouths of ravines, but occur scattered along both banks.

(vii) There are no gorges, big screes, or spurs projecting far out into the river, so that the latter maintains a much straighter course, and there are none of those abrupt S-shaped curves so typical of the Salween.

(viii) Some of the tributaries are of considerable size, and flow quietly into the Yang-tze from wide valley mouths. Nowhere is the basin of this river so narrow as are those of the other two.

Possibly the Yang-tze is nearer its base level of erosion than either of the other rivers with which we are comparing it, and hence is much older. This at least would be in accord with the theory that the watershed between the Mekong and Salween rivers was thrown up at a time subsequent to the elevation of the Yang-tze-Mekong watershed. In spite of the differences mentioned above, one meets with the same rocks in the valley of the Yang-tze as in those of the Mekong and Salween—limestone, granite, and metamorphic rocks.

After leaving the coracle we secured porters from the village and started for the ferry a few miles down the river; meanwhile dusk fell and a stormy night set in, heralded by distant thunder. Just as we embarked in

---

[1] A combination of these two characteristics precludes, as already remarked, the possibility of rope bridges over the Yang-tze.

the scow, the storm burst right overhead, torrents of rain pouring down.   In the darkness the river, now white with foam, was revealed by brilliant flashes of lightning which seemed to fill the valley with red fire, and the thunder roared amongst the mountains.   As the storm had come so it passed, travelling rapidly south-east, and the thunder grew fainter and fainter till it died away altogether.   But for an hour the lightning danced far down the winding valley, like a candle flickering in a draught at the end of a long passage.   A long weary tramp leading the ponies and stumbling over all sorts of obstacles brought us at ten o'clock to the village where we had slept previously, and we continued to Pa-mu-t'ang next day.

On the following morning August 8, we climbed to the summit of the eastern watershed, finding the same lime-stone towers and vast screes crowning the ridge as above A-tun-tsi.   Two lakes I found here, at an altitude of between 16,000 and 17,000 feet, undoubtedly occupied rock-basins, and the belief which had been gradually forcing itself upon me, that the Mekong-Yang-tze watershed had previously been glaciated, became more firmly rooted than ever.

From the summit of the pass (about 17,000 feet) I looked westwards across the Pa-mu-t'ang valley to the rolling hills of the Tibetan plateau, where I caught sight of an extensive lake, bearing rather north and some distance west of Pa-mu-t'ang, but the weather was so thick that I obtained only momentary glimpses of it, and do not know either its position or size.

We returned to Pa-mu-t'ang for lunch and I now sent Kin to A-tun-tsi with instructions to go over some of our old tracks, collect any seeds that were ripe, and mark down all new flowers of any interest, while the rest of our party, namely Gan-ton, the soldier from Batang, and myself, with two baggage ponies, left the A-tun-tsi road and struck off westwards across the plateau by the Jung-lam, the great road that crosses Asia from Peking to Kashmir.   It was miserably cold up here at 14,000 feet, and the rain which had held off for a part of the morning, now came down worse than ever.

At the highest point reached we passed the old boundary stone marking the frontiers of Tibet and China, set up I know not when, for the inscription is all but defaced. The road was also marked by numerous mani pyramids of white crystalline rock, probably quartzose, which must be conspicuous on all but the darkest nights. These mani pyramids consist of long low piles of rock slabs, on each of which is carved the familiar Tibetan prayer "*Om mani padme hum!*" the letters being sometimes beautifully coloured and ornamented. Wooden posts, surmounted by a crescent and a cone, usually crown these cairns, and if it is a very long pile—many extend for hundreds of yards, or there may be strings of them together—there will be several such posts. Though the pyramid is never more than a few feet high, hundreds, nay thousands, of hard stones, painfully carved, go to its formation, and one trembles to think how many hours' work they represent. But what is time to a man who is trying to acquire merit and stifle desire!

In spite of the rain and cold however, the plateau was not altogether dismal, for the bright green grass dotted with the erect racemes of a *Potentilla*, and in parts blue with a Boraginaceous flower, was a pleasant contrast to the barren mountains we had climbed earlier in the day. In the shelter of the mani pyramids the tall blue spikes of larkspurs and other flowers showed up vividly, but the plateau here was quite treeless, and even dwarf shrubs were extremely few and confined to sheltered spots.

In the evening we descended into one of the plateau valleys, and fording a river which rolled down a torrent of red water we reached the miserable village of Lham-da, which, in spite of the black mud a foot deep in its narrow streets, boasted a small lamasery and several substantial houses of mud and stone.

From Lham-da we continued across the heads of several small valleys leading down from the plateau above to the Garthok river, the intervening spurs being well forested on their northern slopes, and the valleys, through which wriggled swift little streams, covered with rich pasture. In the valleys the going was very bad, for the ground was a quagmire owing to the incessant rains and we had to cross numerous swollen streams, but a good solid road led over the spurs, from one of which I caught sight of snow mountains in the west, probably Ta-miu, of which more anon. In the lower valleys, limestone only was visible, and here the monotony of rolling grass-land hills was frequently interrupted by hideously bare scarps, sills, and formless bosses of rock. As we continued in a north-westerly direction however, we came back into the red sandstone country, the strata dipping at high angles, so that curving across the backs of the long low hills they resembled the fleshless ribs of some huge leviathan stretched out to die.

In this corner of Tibet the plateau seems to have a foundation of limestones bent into the form of a basin, or perhaps thrown into a series of folds, the hollows having been subsequently filled up with sandstone, though not to a sufficient depth to overlap the limestone along the rim of the plateau, which consequently appears round the eastern edge, and in every deep valley.

Scattered huts and occasional small villages—at one of which, called Ngu-shi, we stopped for lunch—occur in the valleys, all with large scaffolding frames on which to stack hay and straw standing on the outskirts, and looking for all the world like the beginnings of a London hoarding. Barley is almost the only crop, though a few vegetables such as turnips are grown. The plateau is covered with grass, but the narrow river valleys are well forested with oaks and conifers, and thickets of *Hippophaë rhamnoides* conceal the streams very much as osiers nestle over them in East Anglia.

In the afternoon we reached Phula on the Garthok river, here very much smaller than where we had last seen it at Chia-ni-ting. It was, however, in full flood, deep and swift, the water of a bright red colour from flowing over the sandstone. The post-house at Phula is kept by a Chinaman, as are most of the inns and shops on the Lhasa road. *This* is the real invasion of Tibet, but it has failed for the reason that the women have Tibetanised their husbands. How can the alien from the rich valleys of the Flowery Kingdom maintain his nationality in such a dour land?

All the post-houses were placarded with Imperial Government edicts in Chinese and Tibetan, setting forth

Plate XXVI

View on the grass-land plateau, S.E. Tibet, 13,000 feet

the duties of the inn-keeper, scale of pay, and so on ; and on presenting my *ula* passport I had no difficulty in getting a change of ponies. What was set forth on my passport I cannot say, for even had I been able to read Chinese it was quite illegible ; but attached to it were two small squares of paper, on one of which was depicted in crude outline a yak, on the other a pony or mule, both bearing official stamps ; so that the meaning was sufficiently plain, the pictures speaking for themselves.

In the well-wooded valley of the Garthok river hares were numerous, and not having my gun with me, I amused myself by trying to shoot them with the soldier's rifle, but met with no success though they were ridiculously tame, merely pricking up their ears and running a few yards after each report. A hare for supper would certainly have been a godsend just then, as I had had nothing but eggs and bacon for several days—perhaps that was why I could not shoot straight. The river twisted and twined through the valley in serpentine fashion, sometimes overflowing the grassy flood-plain but often enclosed by cliffs of red earth as much as twenty feet high, river terraces being well defined in places. At nightfall we waded through a foot of mud to the squalid village of Chia-ta-tih only about ten miles short of Garthok, and the four of us were herded together into a single-roomed hut already occupied by a large family. For the first half of the night an old man droned " *Om mani padme hum* " till he finally prayed himself to sleep, and for the second half two babies kept up a continuous coughing and spitting, which was scarcely surprising considering that their dress consisted of one goatskin garment apiece, while the altitude was about 10,000 feet, so that even in summer, owing to the continual rain, the air was very chilly.

Above Chia-ta-tih, which I found next morning to consist of half-a-dozen hovels built of rough logs or of stones insecurely plastered together with mud, and a lamasery with fifteen or twenty priests, the valley begins to open out on to the plateau, and forest ceases. The valley floor, affording the richest grazing ground imaginable for large flocks of yak, sheep, and ponies, reaches a breadth of 300 to 400 yards, and the brilliant green grass is spangled as before with patches of yellow *Pedicularis*, blue Boraginaceae, crimson *Rhinanthus* and other flowers.

Rounding a high rocky bluff, the summit of which was crowned by a temple, we came suddenly upon Garthok, until this moment quite invisible owing to its being built in a depression, so that the flat house-roofs came to be almost flush with the general level of the valley floor. Indeed had it not been for bunches of poles sticking up here and there carrying prayer flags, I should scarcely have seen it at all till I rode into the narrow street, and even then it turned out to be a much bigger place than I had supposed at first glance. According to Captain Rawling's account, villages with important names in western Tibet turn out to be miserable places with half-a-dozen huts, and such for instance is the case with Garthok near the source of the Sutlej, though it is nevertheless an important trading centre in the summer. In eastern Tibet, however, villages which are important are generally also large and the Garthok of Kham province (known to the Chinese as Chianca, by which name I shall in future refer to it, thus avoiding any confusion with the Garthok of western Tibet) is almost as big as Batang, boasting I suppose quite 200 families, a lamasery with over a hundred priests, an official yamen with a small garrison, and no less than four schools!

The moral of all this is that trade between China and Tibet is in a far more flourishing condition than trade between India and Tibet, and that it is the Chinese who have made eastern Tibet what it is. Nor is it to be thought that all this has taken place only since 1905, for the Chinese have been in eastern Tibet for more than a century, gradually building up the present prosperity, though it is well known that the events of 1905 gave an impetus to their activity.

At the time of my visit to Chianca the official and garrison were away, for throughout the summer Chinese and Tibetans had been fighting in the Pomed country a fortnight's journey west of Chianca, and though local rumour said that the Chinese were losing heavily, such statements must be taken for what they are worth. Many of the houses in the village are built several feet below the level of the cobbled street, and being only a single story high, one looks right over the first floor. Often there is only a single room, which being partly underground is dark and dungeon-like. The numerous small shops are kept by Chinese and semi-Chinese, but the population of the village is mostly Tibetan, at least in appearance. The women put black grease on their foreheads and cheeks, a cosmetic which is supposed to prevent the skin cracking in the cold winds, though Huc, describing the origin of this habit, gives a much less prosaic reason for the black grease; but it is not beautiful. They wear their hair in innumerable tiny pig-tails which are collected together at the waist and woven into a single artificial plait hanging down to the ground ; or the entire contrivance may then be wound round the head.

As usual, the official inn is kept by a Chinaman, who however is Tibetan in everything but birth. He showed me some wounds, the result of 800 blows recently inflicted by the official because the inn was not in a fit state of repair for government officials, and asked me for something to rub into them.

As for the schools at which all children are bound to attend, I visited one, and found the children singing Chinese sounds out of a book of characters. Boys and girls were in separate rooms and the Chinese pedagogue in a third room, I suppose waiting till the children had learned a certain number of characters, when they would be called upon to repeat them without the book, or recognise a given character at sight. It is really astonishing what the Chinese have done as regards establishing schools in the smallest villages, though whether the idea of making all Tibetan children learn to speak and read Chinese is meeting with the success anticipated, is another matter.

I had scarcely settled down in the village for the day when a Chinaman came and begged me to visit his son who on the previous day had been bitten by a dog. I found the child, who was only six years of age, sitting quietly in the room, with a nasty wound in the right lower jaw, now a horrible mess of clotted blood and red mud which had been plastered on to staunch the flow of blood. It took half an hour to clean it with warm water and a lancet, for I had literally to cut away the scabs, an operation which the patient stood with great fortitude ; but when it came to cauterising the wound without stint there were shrieks and struggles. However I was adamant and thoroughly burnt out the surface flesh, after which I bandaged the wound and made the little fellow as comfortable as possible. I wanted to try my hand at putting in stitches but the father thought the boy had stood enough, so I desisted, at the same time pointing out that it would leave a deeper scar.

In return for my work I asked the father to collect for me and send to Batang the seeds of certain flowers I showed him just outside the village, about five plants in all. This he promised to do, but the seeds never arrived at Batang, and distrustful of Chinese gratitude, I decided that never again would I attend a Chinaman unless I knew him. After all, I had used up all my caustic on that boy, and it might have been awkward if I myself had been subsequently bitten by a dog !

Another patient who came to me in Chianca was the Tussu, an old man of sixty, who showed me some nasty sores on his wrists and ankles, filthy with mud and pus. I simply washed them and dusted them over with iodoform, for which kindness the Tussu sent me several rotten eggs and a big bowl of excellent milk.

Later in the afternoon I went up the valley and scrambled about on the sandstone cliffs, now gay with flowers such as *Meconopsis Wardii*, several Labiatae, species of *Campanula*, *Pedicularis*, and so on, making a very striking show against the red rock, in strange contrast to the fields of barley down below which were suffering badly from smut. Meanwhile Gan-ton had made arrangements for the continuation of the journey next day to Samba-dhuka on the Mekong, as I wished to return by a different route. My Chinese soldier had now to return to Batang, and I was given instead a Tibetan soldier. I have already remarked that the official was away, and had I wished, there was nothing to prevent my continuing along the road to Lhasa, distant about five weeks' journey over extremely difficult country ; but time was getting on and I was due back in A-tun-tsi if I wished to make the most of the season in that region.

CHAPTER XI

THE WONDERFUL MEKONG

LOOKING back as we rode up the valley on the following morning I quickly realised how it was that the forest stopped short here at 11,000 feet, for the low undulating hills offered no protection whatever to the fierce winds which sweep down from the high plateau and rush through the narrowing jaws of these valleys. Grass-land alone could withstand such a scourge, and even so the rib-like sills of harder rock stand out bare and barren on the steeper slopes.

To the north the sky was as black as ink, and we had no sooner turned off westwards just above the village—the main road to Lhasa continuing northwards—than a drenching rain-storm was upon us. Several Tibetan horsemen, wearing long red cloaks of rich cloth and broad-brimmed felt hats with red crowns—altogether a quaint garb—passed us, but there was little traffic on the road. At length we reached the plateau-watershed, our direction being about S.S.W.; in the distance immense patches of blue indicated the brilliant Eritrichium, and *Meconopsis Wardii* was abundant.

We had scarcely begun the descent of the richly-forested western slope than there burst upon us the most terrific rain-storm I have ever experienced either within or without the tropics. The noise made by this deluge was extraordinary; in an incredibly short space of time rivers were rolling down the hill side, the ponies could hardly keep their feet on the slippery turf, and we were all drenched to the skin and so numbed by the cold that I could with difficulty clutch the sodden reins in my swollen fingers. It was useless to seek shelter beneath the trees, for the wind drove the rain straight at us, but it was equally impossible to face it; the loads became loose, the men lost their tempers, and I began to regret having ever adventured myself in these storm-swept Tibetan uplands.

This was obviously the rainy side of the plateau, for the thick forests of fir extending above 13,000 feet told their story as plainly as did the much dissected ridges, the curious isolated cones of earth, and the wide-mouthed valleys with their fan-shaped loads of detritus. These latter were short and broad, forming equilateral triangles, as did the valleys from which they had been shot out. Lower down, the valley floor was littered with large boulders and gravel through which the torrent, now grown to considerable size, cut its way, sometimes flowing in a deep ravine. No doubt the heavy summer rainfall of the plateau country is due not so much to the rain-bearing winds from the S.W., which we have already shown to be stopped by the Mekong-Salween divide, but to the local hot winds blowing up the Mekong rift at last unloading on the high cold plateau the moisture they have gradually accumulated.

The Mekong gorges are here deeper, narrower, and more arid than ever, but this local gale cannot blow day after day for several months through all these miles of gorges without sooner or later throwing down some moisture. A point must be reached at which the pressure becomes such that the hot air is forced to rise and spread out, as it seems to do north of A-tun-tsi, where the Mekong swings away to the north-west and a long arm of the plateau is thrust down between it and the Yang-tze. It is natural to suppose that the broad expanse of high plateau would condense more rain than does the higher but extremely narrow ridge further south, and it is this difference of rainfall that we want to account for.

After condensing most of their moisture on the Mekong-Salween divide, the prevailing winds sweep on over the open plateau, which for that reason—namely, the exposed nature of the country at this high altitude—can support nothing but grass-land. By the time the high limestone ridge overlooking the Yang-tze is reached there is scarcely any moisture left, so that, as usual, barren screes prevail above 16,000 feet.

Thus in the plateau region, starting from the Mekong

where it begins to separate from the Yang-tze, we get, first, the arid region in the deep gutter itself; secondly, going eastwards and ascending, forest; thirdly, grass-land; and finally above the Yang-tze, scree; which is exactly the sequence in the horizontal plane which we find in vertical extension on the Mekong-Yang-tze divide further south. That is to say, the grass-land plateau is simply the equivalent of the alpine grass-land to which we have already referred as a prominent feature of that ridge.

Under the miserable conditions I described before this digression we resumed our march after a halt to readjust the loads.

Below the gravel region we came again upon cliffs of limestone, and of red sandstone conglomerate with water-worn pebbles embedded in a hard siliceous matrix. Approaching signs of habitation we presently came upon an image of Buddha, extravagant in design and execution, painted in crude colours on a smooth limestone cliff: close by was the village of La-wu-rah, amidst terraces of barley and buckwheat.

Away to the west, in Tsa-rüng on the other side of the Mekong, snow mountains appeared, phantom-like amongst the clouds. This range, of which I was destined to get a much better view later, is known to the Tibetans as Ta-miu, and is evidently the northward extension of the K'a-gur-pu ridge on the main Mekong-Salween watershed, but whether it is continuously above the snow-line from K'a-gur-pu onwards is a point I could not decide.

As we descended, the valley took on more and more the character of the now familiar arid region. Granite and metamorphic rocks re-appeared, and stunted bushes of *Sophora viciifolia* took the place of forest. Presently we crossed the terracotta-coloured stream, here rather broader and shallower than above, but though we could hear the boulders grating and rattling against each other as the flood swept them along, and the ponies were a little shy of entering the water, we had no real difficulty in crossing. Later our cavalcade floundered into a quagmire in the valley bottom, and further on we found the cliff path blocked by two enormous boulders which had rolled down the mountain side. Happily the largest of them was nicely

balanced or we could never have stirred it, but the combined efforts of five men fortunately sufficed to send it
hurtling down the slope and we were able to pass.

We had left the worst of the rain behind now and it
grew clearer as night drew in. At the village of Wa-ka-tih
we changed ponies for the second time, and in view of the
fact that it was getting dark and we had still a long way to
go, I told Gan-ton to see to it that we did not waste time
changing ponies again.

Across the valley was a village built on a steep slope,
every house supported on piles. Moreover the houses
were built close together, so that in the dusk the place
looked for all the world like a Malay village. This method
of raising the houses on piles is common amongst the
poorer people of S.E. Tibet as well as amongst the Lutzu
tribe, though I do not think the people who affect it are
genuine Tibetans. With their dug-outs, fishing-nets, and
pile dwellings the Lutzu remind one forcibly of the Malays,
and indeed there may well have been some connection
between the tribes of these regions and a seafaring folk;
otherwise whence do the Lissu women get their cowries?

At the next village, called Ndu-er, all the women
and children flocked out to have a look at the stranger,
dogs barked, and everyone talked and shouted at the same
time; then two of the men, in spite of the soldier's orders,
insisted on unloading one of the ponies, whereupon he
jumped down in a great rage, and picking up two rocks
each the size of a quartern loaf, flung them with all his
force at the offender standing but a few feet from him.
The first missed its mark, nearly brained a child standing
just behind, and ricochetted off the ground on to the hind
quarters of a dog, who went off yelping; but the second
one caught the victim fair and square amidships, luckily
just where the sleeves of his gown were tied round the
waist, making a thick pad. He doubled up like a shot
rabbit, but apparently no serious damage was done; still,
I should not like to have been the recipient of the missile.
I now interfered and through Gan-ton put a stop to
reprisals. Every man carries a sword here and we should
have seen them bared next minute.

The incident cost me some money on the principle of

the Workmen's Compensation Act, for my soldier with true oriental despotism had warned the people in one village that if they did not do what I told them they would be beaten! They are great swashbucklers, these Tibetan soldiers, thriving on reflected glory; and though my guide was only anxious to please me, he was evidently a man of both rank and authority, for every villager cringed before him. As long as the people are quiet, such a reputation as he had given me does no harm, but should they be restless, it might easily occur to them that after all I was alone. In the present instance therefore I tempered justice with mercy and hard cash.

At nightfall we climbed the high spur guarding the valley mouth and found ourselves on a narrow path with a fearful chasm below us, where a long ribbon of water caught the last dying light, which gleamed irregularly on swirling current and racing eddy, while there floated up to us out of the darkness the hollow booming of the restless river. It was the Mekong thundering southwards through the gorges. I have already said that the Mekong is the smallest of the three rivers, having neither the tremendous current of the Salween nor the great breadth of the Yang-tze. Yet when I heard it at night pouring forth from that deep wound in the Tibetan mountains I thought it the grandest river of them all.

The path down the river, though well made, was narrow, and in the gloom the precipices on the one hand, and towering screes stretching up and up out of sight on the other, made riding uncomfortable, so I walked. There was some doubt as to whether we should be able to reach Samba-dhuka, our destination on the Mekong, at all, since the crossing of a big torrent in the semi-darkness was no easy matter. However we presently came upon the stream and crossed in safety, the men wading through waist deep and the ponies struggling across with the water almost up to their girths. About nine o'clock we suddenly came upon Samba-dhuka which, as I found next day, consisted of some twenty or thirty wooden huts, hidden in a mountain alcove, on a boulder-strewn alluvial fan with broad terraces of buckwheat. Everything was very still, but the men approached giving the Tibetan call, and in a twinkling dogs were barking, people were out with flaming torches, and we were made welcome for the night. I was very tired, for we had been ten hours in the saddle, drenched most of the time, but the air down here was warm and balmy, and as the moon rose over the mountains lighting up the thin river mist, I slept soundly again.

Early on the following morning, while the men were getting ready, I went down to look at the Mekong.

Just here the river presented an extraordinary spectacle. For a quarter of a mile it flowed between fluted walls of limestone, not more than fifty feet apart and perhaps a hundred feet high, and looking down from the cliff on to the red water writhing in this confined sword-cut below gave one some idea of the irresistible power of the river. It was impossible to escape the conviction that the river itself had sawn its way down between these cliffs to its present level, especially as there were traces of another similar wall at a higher level. The depth of water piled up in this narrow space must be tremendous, and considering the immense volume that was coming down and the quite puny size of the torrents which flow from the dividing ridges, really distant only a few miles to east and west, I am inclined to think that the Mekong rises further up in Tibet than is generally supposed. For where the Mekong-Yang-tze ridge spreads out into the plateau, it is not the former river which drains the wide expanse of rain-swept grass-land—at least not here, whatever it may do further north.

Across this chasm were slung rope bridges, though I should have shied at crossing by them, and on the other side a mountain road led away across Ta-miu into the heart of Tsa-rüng; but there was no road down the Mekong, and no road up the river either, above the valley we had come down. However, there was luckily no occasion to retrace our steps to Chianca.

Masses of light cloud hung about the valley, and a heavy dew had settled on everything, but no sooner had the sun appeared over the ridge than this was dispersed like magic, blue sky appeared overhead, and the ravine began to heat up for its daily roasting, though columns of cumulus were already towering up in gigantic puffs from

Plate XXVII

Two views of the Mekong at Samba-dhuka, showing fluted limestone cliffs
and rope bridge

the mountains to the west. Still it looked as if we were going to have a glorious day.

Our route lay in an E.N.E. direction up the bed of the torrent we had with some difficulty crossed on the previous evening, and now we had to cross it not once, but scores of times; in some spots it looked a most formidable undertaking, but as we ascended and the stream divided the crossings of course became easier.

The main geological features around us were gravel cliffs, the familiar red sandstone of the plateau, and earthtables, each consisting of a column of earth capped by a flat boulder.

At Du-bas, the first and last village in this gulley, we changed ponies and set out in quest of a new route over the watershed.

Higher up in the forest, above the small cultivated plots, we found that the torrent, here ploughing its way through a deep trench cut out of red gravel, had recently come down in tremendous flood, spreading deep deposits of semi-liquid gravel everywhere, and through these abominable quagmires the ponies had to flounder knee deep; they got thoroughly scared sometimes, and indeed it was most unpleasant.

Forests of fir and of oak marked the shady and the exposed sides of the valley respectively, but down by the stream we rode through groves of poplar, willow, birch, and numerous shrubs to the pass, whence we looked down into a second valley, the separate bands of fir and oak forest being here very conspicuous. So dense was the vegetation that one of the loads was wrenched from the saddle as we brushed through, and after that came marshy ground and thickets of willow, till finally climbing up between really magnificent fir trees, we reached the main pass and emerged quite suddenly on to the grassland plateau once more.

So far the weather had kept fine, though by this time we were well up in the mountains; but no sooner did we get out on to the open plateau than I saw we were in for trouble. A nasty raw wind blew in violent gusts, and right in front of us a great black ridge of cloud hung low over the hills. We had scarcely turned southwards down the valley than the storm burst upon us with amazing fury. The whole earth seemed to rock to the thunder-claps, as the echoes tumbled from side to side amongst the hills; the hail lashed into the short turf with a sharp hissing sound, and drummed on our hats and cloaks. Rills boiled up into streams in an instant and came frothing down the grassy slopes. In less than half an hour the valley was wrapped in a shroud of hail more than an inch deep, and looking at those white mountains so bleak and bare, from which lower down a few black clumps of fir trees stretched their grim spires up towards the leaden sky, I thought that winter had already breathed again over the grass-land plateau of dreary Tibet. Yet it was scarcely mid-August.

Wintry enough it looked and felt as I rode behind, following the deep imprints stamped hard in the carpet of hail. Now the ponies cantered and galloped to keep warm, turning their heads sideways to avoid the fusillade of ice, crunching and slipping on the slopes, shying at the dark ribbons of water which every few yards opened up before them, sliding down the muddy banks of the streams, plunging up to their girths in the swollen river, and stumbling in the concealed pica-burrows which honey-combed the ground in all directions.

Presently we came to signs of habitation—small fenced-in portions of the grass-land where barley was ripening, and scaffold frames had already been erected for stacking straw. But there were no houses. Lower down, where the trees began, we passed a few yak herders huddled under a tree, their long coarse cloaks wrapped closely round them, a black smoking fire their sole comfort. Wet and cold as we were, our plight was far less miserable than theirs, though they were doubtless quite happy. To keep up our spirits Gan-ton and I sang songs as we rode along; after all there was a fine feeling of freedom and irresponsi-bility while in the company of these happy-go-lucky re-sourceful Tibetans, and when a man feels in first-rate health, a few hardships only make him more conscious of his fitness.

At dusk we reached the last house in the valley, a big solid two-storied building, and entering it we were at once made welcome; indeed the good people built up such a furnace in the room allotted to me that some of my clothes were burned and I was soon driven outside by the heat, and compelled to have supper on the roof. All night long it poured with rain, and shortly after starting on the following morning we were as wet as ever.

A few miles below, our stream entered the Chianca river not far south of Phula, and here our soldier left us to return to his home, while we, with another Tibetan lord as escort, followed down the Chianca river in a southerly direction.

Gradually the valley deepened and high sandstone bluffs appeared here and there, one of which was decorated with a number of ancient Buddhist carvings of unknown antiquity; but unlike the high plateau valleys which only offer facilities for grazing, and are occupied by a nomadic pastoral people, there was here plenty of cultivation on the steep slopes.

My second soldier proved willing to go to even greater lengths in my service than the first had done, for a certain woman having made some trouble about changing ponies, he rode straight at her with uplifted whip, prepared to lay it across her bare shoulders, a chastisement which she escaped by dashing into the house.

For the second time I interfered, and dismounting, took the whip from him and threatened to beat him with it if he did not behave—though I am bound to confess that this was more because I wanted the whip myself than because I disapproved of his action.

On general principles I consider it neither expedient nor of the slightest use for a traveller to interfere blindly with native customs, and I am sure the woman was far more astonished to see me tackle the headman than she would have been to feel the whip across her shoulders. Also I doubtless made myself very unpopular with the other villagers, who resent any form of interference with their ruling class. However I secured the whip, a very nice leather one, though I gave the man a rupee for it afterwards.

Next day we recrossed the watershed between the Chianca river and the Mekong; quite a short climb, for we reached the latter river early in the afternoon. The weather

was fine, and from the summit we had an extensive view of
Ta-miu, though the high peaks were so buried in cloud that
I could only make out one glacier with any certainty.   It
is, however, a fine range of snow-clad mountains, but it was
again impossible to see whether it extended southwards to
join K'a-gur-pu, or whether these two *massifs* are distinct
elevations of the main watershed.   Personally I believe
that, far back between the Mekong and the Salween, a
continuous chain of snowy peaks does extend from K'a-
gur-pu to Ta-miu and so northwards into Tibet; but it was
impossible to prove this while in the deep Mekong gorge
itself.   Geographically it is not a matter of any importance,
but botanically speaking, it might be.

Though we reached the Mekong soon after two o'clock,
a delay of nearly four hours occurred before we could secure
transport animals, and the journey in darkness down that
perilous Mekong road proved most exhilarating; there was
no moon, but millions of stars shone in the clear sky, and
it was beautifully warm.   High cliffs of gravel and rubble,
capped by enormous overhanging boulders in most insecure
positions, were a feature of the valley here, with the usual
dangerous screes, deep gullies, and broken road.

It being impossible to reach Y'a-k'a-lo in decent time
that night, we stopped at ten o'clock, continuing early on
the following morning, and at Y'a-k'a-lo we got into difficulties with the official, who at first refused to supply us
with transport animals.   I went to see him myself however,
and he received me with every mark of respect, at the
same time pointing out that as I had come by an unauthorised route from Chianca, I must return to the Batang
road.   The Mekong road to A-tun-tsi was very unsafe, he
told me, for an American traveller had killed a Tibetan
there several years previously, and he was consequently
afraid for my life.   Moreover he seemed terribly upset
about Captain Bailey's exploit, which had got him into
serious trouble with the Viceroy of Yunnan.   However I
had to get back, and eventually the official compromised ;
he would give me ponies if I would sign a statement
exculpating him from all responsibility and saying that he
did not know where I was going !

Of course I complied, as it did not matter a scrap to

Plate XXVIII

A Wayside Temple in the Mekong Valley below Y'a-k'a-lo

Crossing a torrent in S.E. Tibet

me what I wrote, and having written out the document, in English of course, I left them to write the Chinese version as they liked, since I should have been none the wiser had I seen it.   Had anything happened to me before reaching A-tun-tsi, I can well imagine how an astute critic would have pointed out that the official's obligation hardly ended with the signing of a document in which neither contracting party understood the writing of the other!   Nor does it seem to have occurred to the official that I might have written anything I liked, even to a serious indictment of himself.   Of course I played the game, but I am glad for his sake that he never had reason to show the document to the Viceroy, since it was boldly stated that I was returning from Y'a-k'a-lo to A-tun-tsi by the small road, unknown to the local officials!   I imagine the first comment the Viceroy would have made would be that it was his business to know!   The official refused me an escort on the ground that the journey was quite unofficial, and for that at least I was devoutly thankful; but he had no right to stop me, and he knew it, in support of which statement I could quote no less an authority than the late Warden of the Marches and Viceroy of Ssu-chuan, His Excellency Chao. And being faint-hearted and fearful, he compromised.

There was nothing remarkable about the three days' journey down the Mekong except the extraordinary height of the road above the river, which gave me a vivid idea of the depth of this amazing gutter, for the valley walls still towered high above us.   Sometimes we must have been 3000 feet above the water, so that we could look down on to great rapids and yet hear no sound.   No journey could be more appalling in its wearisome monotony than that through these arid gorges ; climbing up and down over the endless spurs, sweeping in round the endless gullies, we would make good but a few miles in a straight line after hours of riding.

The morning of August 16 saw the mountains on either side covered with snow, but it melted again during the day, though two or three snowy peaks were now visible at the heads of gullies to the west, again suggesting the continuity of the K'a-gur-pu snowy range northwards to Ta-miu.   The plants were those of the arid region, though so high up were we, sometimes over 10,000 feet, that there was a considerable variety of shrubs.

On the 17th we set out very early, as I wished to reach Adong that night, and at midday I left the caravan and rode on ahead, confident that the men would follow on till they overtook me. Unfortunately I did not realise how far we still were from Adong, and the most execrable part of the road was still to come. In places it was horribly dangerous, and for miles two ponies could not have passed each other by any possible manoeuvre save that of leap frog. I was thankful we met none. To make matters worse my pony threw me and then ran away, and I had great difficulty in securing him, though happily he could not leave the path. The long hot days in the saddle had made me feel dazed and drowsy, and being thoroughly sick of the sight of the Mekong, I paid little attention to the scenery, though one big cataract we passed was particularly fine.

It was dusk when I reached the dark gorge leading up from the river to Adong, and in the waning light nothing could have exceeded the grandeur of the scenery here— towering limestone cliffs with scattered clinging fir trees, the torrent pouring over the rocky precipice in a mighty cascade which filled the gorge with thunder, and looking back, the vast icy pyramid of K'a-gur-pu, pallid and almost unreal against that velvet sky, blocking up the mouth of the ravine.

Long before I reached Adong it was pitch dark, and not knowing the road, I was forced to dismount and lead my pony. Once we stopped instinctively; we were not on the path at all, but on the brink of a precipice. However, eventually we reached the scattered village, and selecting a familiar-looking house as likely to be the one we had stayed at previously, I made an assault upon the door.

"Hullo there!" I shouted in Tibetan, and immediately came the answer, "What do you want?"

Unfortunately I had now come to the end of my Tibetan vocabulary, and had to go on in Chinese. "Open the door," I said; "I want to stay here to-night, and will give you money."

But I might as well have talked English, for all I was

Plate XXIX

K·a-gur-pu, from an altitude of 12,000 feet, near A-tun-tsi,
looking west

understood. They went on at me in Tibetan while I kicked the heavy wooden door and cursed them heartily; finally I put my shoulder against it and tried to break it open. But it was too strong and yielded not an inch.

Meanwhile there was great excitement within, people running about with torches and shouting all kinds of things, but not a man would show himself, and I could not climb the wall. Had they seen me, all might have been well, for I was known in the village, but apparently it did not occur to them who it was, and fearing a robber ruse they would neither open the door nor expose themselves.

As I kicked viciously at the door, there suddenly fell a great silence upon the household, which I presently noticed and took to heart; it was the more impressive on that dark night after the recent racket and the lurid glare of pine torches. I thought : " They have gone for their guns; I had better get out of the way," so with one final effort I dashed against the door, which cracked, but would not yield, and to the accompaniment of a long Tibetan curse, I ran away into the darkness laughing.

But my pony and I were now dead tired and it was with a sigh of relief that I eventually found the Tussu's house, where we had as a matter of fact stayed on our way to Batang. Here at least I would be well received, so I set about knocking up the household.

There was a light burning in the big lonely house, but shout as I would, I could get no response. Finally tying up my pony, I scaled the wall; something was shaking itself in the darkness below, and I dropped down almost on the top of a big mastiff, who had roused himself, and at once set up a tremendous baying, so that I narrowly escaped with a whole skin. Before the household could be alarmed I had climbed up to a window, and crawling through I dropped to the floor and entered the big household kitchen, dashing straight into the midst of a crowd of children playing some game uncommonly like hide-and-seek; it was at any rate equally noisy, which accounted for their disregard of my entreaties for admission.

A panic immediately ensued, and everybody ran away shrieking except a few elders, who stood in the middle of the room staring at me with open mouths. Happily there was a Chinaman sitting by the fire and to him I explained my position, quickly reassuring everybody.

They got me some tea and *tsamba*, sent a boy to look after my pony, and generally did what they could to make me comfortable, which wasn't much; for my goods not having arrived, I was forced to sleep on the floor with my saddle for a pillow and the saddle-cloth for a blanket. I do not remember ever before to have been so full of aches as I was that night after fourteen hours riding and walking—till the following morning!

Having partaken of a little more Tibetan fare at five o'clock next morning, I set out for A-tun-tsi and had reached the other end of the village, a couple of miles down the valley, when I fell in with Gan-ton and my caravan, who had arrived at eleven o'clock on the previous night, and hearing no word of me, had taken refuge in the first house they came to thinking that I had gone on to A-tun-tsi. I called a halt at once and made Gan-ton prepare me a square meal with lots of fresh yak milk and new-laid eggs, after which we set out for A-tun-tsi, reaching our base camp at midday, having been absent just over three weeks.

For the excitements of that hurried journey to Batang and thence to Chianca in S.E. Tibet I had to thank the French priest at Tsu-kou with his amazing story from Lhasa, which strange to say he still persisted in, even after I had effectually exploded it!

CHAPTER XII

MOUNTAIN AND MONASTERY; A SECOND JOURNEY TO THE YANG-TZE

FROM August 19 to September 16 we remained at A-tun-tsi, except for three days in camp on the eastern range.

Day after day we went over the old climbs, finding many of the spring flowers in seed; on the other hand the summer flora at high altitudes was now at its best, great numbers of gentians being in bloom on the grass-land while the limestone rocks were gay with bunches of yellow saxifrage. Lower down was a wealth of Labiatae, mostly herbs of rank growth, the majority of the common British genera being represented and, like the gentians, saxifrages, and other genera rich in species, flowering simultaneously. There were also many species of *Corydalis*, both in the alpine region and at intermediate altitudes, some of them growing as shade plants, some on the open screes at 16,000 feet, while one, with dense spikes of brilliant yellow flowers, was an aquatic.

On the dry rocky slopes exposed to both sun and wind, just below A-tun-tsi, I found a purple-flowered *Morina* (Dipsaceae), white-flowered specimens also being frequent, and growing on a limestone cliff at 13,000 feet was a small *Pinguicula*, but unfortunately the flowers were over. I also came across the pretty twining *Codonopsis convolvulacea* with large mauve flowers, and at 15,000 feet was an erect species of this same genus with pendent bell-shaped corollas of a dirty flesh colour prettily veined with crimson inside, but having a most abominable odour. The genus *Codonopsis* was represented by yet a third example, which occurred at

11,000 feet; this also was a twiner with flowers like the last-mentioned species and the same disgusting odour.

One day we made the complete circuit of the high mountainous ridge to the west, Kin having discovered what he considered a practicable route. However, before I knew what was coming he had led me to the brink of a clear drop of some thirty feet high, with a steeply-shelving scree below, and down this he coolly climbed gun in hand, though what he held on to puzzled me. In my descent I stuck half way down in fear of my life; while Kin, standing on the screes below, encouraged me with shouts of "Don't be afraid! Don't be afraid!" Finally I got down, feeling very uncomfortable, whereupon Kin remarked quite casually: "I was afraid when I came here a few days ago that if I fell no one would find me!" for he had prospected this hazardous route alone.

The Chinaman does not seem to be troubled with nerves in these matters, not, I think, because he has any less instinctive objection to death or mutilation than the generality of mankind, but because it does not readily occur to him that he might fall off a ledge of rock a hundred feet high, any more than he might if it was only two feet above the level.

In spite of frequent showers the weather on the whole remained very good till the end of the month, though heavy storms regularly passed over the Mekong valley from K'a-gur-pu to Pei-ma-shan, distant about five miles from A-tun-tsi as the crow flies. Frequently we heard thunder from that direction, and one night a heavy thunderstorm with brilliant flashes of lightning and drenching rain passed over the village. Towards evening, when bright sunshine prevailed at A-tun-tsi, we would sometimes see vivid rainbows thrown against the heavy blue-black skies over Pei-ma-shan, and as early as August 28, the lower slopes of the mountain right down to the pass at 15,800 feet were white with snow, but it did not last out the day. This was the first snow we saw on the Mekong-Yang-tze divide, though already it must have been snowing frequently at this altitude on the Mekong-Salween divide.

One day Sung came to me weeping and asked if he might go back to Tali. On pressing him for particulars of the trouble, he said that he had no money with which to buy bed-clothes now that winter was coming on, and that he was cold at night and was always feeling ill; moreover he had heard it said that I considered him lazy—which was quite true. However, I had no desire to lose him so near the end, and it was a matter which could easily be put right, so after rowing him for not coming to me at once when he felt ill I gave him some medicine, his wages more than a month in advance, and some extra money for bedding, lending him my rug in the meantime.

A short time afterwards I discovered that he spent nearly half his wages every month on Chinese spirit, which he drank daily, and so far was he from wishing to return that when I taxed him with this vice and told him he could return to Tali if he wished, he did not want to go! To a certain extent I felt duped in the matter, for though I had kept an eye on my men both as regards their own wants and my interests, and had at the outset told Kin to report to me as soon as either of them felt ill or required anything, I was quite ignorant of Sung's habits. However he gained nothing by it, for I eventually deducted the extra silver I had given him from his next month's wages, and though I gave him a small present when he left my service, I did not reward him as I rewarded Kin's faithfulness when I parted from him in Bhamo. Sung was a good cook, but his utility stopped short at that.

On September 1 we went into camp in one of the valleys above A-tun-tsi at an altitude of 13,000 feet. Next day the porters returned to A-tun-tsi, taking with them the ponies; and leaving Ah-poh to guard the camp, Kin and I ascended the first hanging valley to the alpine grass-land above, finding several species of *Meconopsis* and *Primula* in seed, as well as many plants in flower, chiefly saxifrages, gentians, dwarf aconite, and larkspur.

It is interesting to note that the seeds of a large proportion of these high alpines living at from 16,000 to 18,000 feet are adapted for wind distribution. For example those of the dwarf rhododendrons are winged, those of the saxifrages and gentians extremely small and light, those of the numerous scree composites provided with the usual pappus, and so on. It is particularly noticeable at these altitudes in the case of precipice plants, which may be compared to epiphytes. Considering that the winds blow up these valleys, the fact that wind-borne seeds should reach such high altitudes and gradually colonise ground which may have been slowly exposed owing to the retreat of glaciers, is quite natural, and I have frequently watched seeds of Compositae being whirled up a mountain side 12,000 feet above sea-level. On the other hand the mere possession of seeds capable of being carried by the wind does not enable a plant to establish itself on these inhospitable mountains, for neither *Epilobium* nor *Clematis*, to mention only two genera which are common at 12,000 feet, has representatives in the true alpine region. Nor must it be forgotten that the high alpine flora is a north temperate one, and has undoubtedly come from the north, though growing here in the latitude of Cairo; so that the south winds which blow up the deep main valleys as already described cannot have added anything to the composition of the flora.

In the night it rained heavily and turned very cold, so I went out and brought my faithful mastiff Ah-poh inside, to his great joy. In the morning we ascended the main valley and branched off to a pass at the head of another hanging valley, about 16,000 feet above sea-level. From here we had a fine view of the rocky summit of the ridge under deep snow. The Rün-tsi-la, as the pass at the head of the main valley is called, was still a long way above us, and it was this pass to the Yang-tze that I was most anxious to cross. Enormous piles of angular rock fragments surrounded a small lake at the foot of this hanging valley, and here I shot a young marmot. In the evening one of Chao's soldiers turned up with the porters, saying we had to return to A-tun-tsi, and as we had got all our seeds in and it was extremely uncomfortable in camp in these constant rains, I acquiesced, and on September 4 we went down.

The village was gay and noisy now, for it was the Mohammedan New Year, and no sooner was that over than we had a Chinese festival, the narrow street being decorated with large lanterns, and *ho-p'ans* of joss-paper burnt outside the houses to the accompaniment of gongs and drums.

Plate XXX

Screes above A-tun-tsi on the Mekong-Yang-tze Divide, 16,000 feet

Meanwhile the rains, which should have come in August and been over before this, continued with unabated vigour, and climbing the mountains in the gales which frequently blew was astonishingly cold work. Sometimes I would wake up in the morning and find my room flooded, water dripping through the roof in a dozen places, my blankets, plants, seeds, books, and everything else wet.

One evening I was in the forest later than usual, and as dusk fell the gloom greatly increased owing to the clouds which had descended into the valley and enveloped everything in a cold drizzle. Suddenly I heard a snorting noise not far away, and peering through the trees I saw a large black animal advancing towards me with an odd gait, his nose close to the ground. At that moment he reared up on his hind legs not a dozen yards from me, and I saw that it was a black bear, about four feet high as he stood. I was too astonished to do anything but stand and gape at him, as for some seconds he stared at me ; then with a particularly loud snort of disgust he dropped down and shuffled rapidly away into the forest. It was a surprise to me to see a bear within half-an-hour's walk of A-tun-tsi, and having no weapon of any kind I was glad he had run away, though annoyed that I had not been able to bag him. The black bear must be fairly common in these mountains, for all the tribes in the Salween and Mekong valleys use arrow-cases made of bear-skin, just as they have saddle-bags made from the skin of the 'precipice sheep,' as the Tibetans call it, though what this animal is I am unable to say, never having secured one. Shooting bears with a cross-bow sounds exciting.

This was not the only big game we came across, for besides Kin's leopard, I twice saw deer above A-tun-tsi, and we had watched precipice sheep scrambling up the limestone cliffs in the Salween valley. Sometimes Ah-poh would stop suddenly on the edge of the forest and bark furiously for some minutes, turning round now and again to look at us, and again turning his attention to something we could not see, reluctant to leave the spot. On one such occasion we heard a heavy body crashing through the thick scrub. The hunter ought to get some good sport in this region with the above-named animals and takin (*Budorcas*) at the headwaters of the Irrawaddy. I once saw the

horns of this animal in a Lutzu hut.   There are also musk
deer, and I dare say other things besides.

Numbers of pilgrims passed through A-tun-tsi every
day on their way to Doker-la—long processions in single
file, men and women, with packs on their backs, and
bamboo wands, each decorated with a sprig of evergreen in their hands.   I now saw for the first time one of
those extraordinary people of whom I had often read, who
proceed by measuring their length on the ground over the
entire distance, thus acquiring a vast amount of merit.
He was a ragged-looking man, dirty and ill-kempt, as well
he might be, with a leather apron over his long cloak, and
his hands thrust through the straps of flat wooden clogs,
like Japanese sandals.   Standing up with his arms by his
side, he clapped the clogs together in front of him
once, twice, then slowly raised them above his head, and
clapping them together a third time, stretched himself at full
length on the ground with his arms straight out in front of
him.   Mumbling a prayer he again clapped, made a mark
on the ground at the full stretch of his arms, and rose to his
feet.   Then he solemnly walked forward three steps to the
mark he had made, and repeated the performance ; and so
the weary journey went on.

Who but a Tibetan, taking no heed of time, consumed
with zeal for a religion which preaches self-effacement, could
devise such a method for acquiring merit !   Perhaps he was
travelling thus to Doker-la, a journey which would take him
weeks to accomplish.

On September 13 the ex-official Hsia-fu, who had as a
matter of fact been dismissed for embezzling government
money, left the village, his influence having obtained for
him another post elsewhere.   Thus was he rehabilitated
in the eyes of the population, and Chinese politeness demanded that his previous pecadillos should be ignored as
though they had never taken place.   Hsia-fu once more
held official rank, and convention had to be observed.
Though degraded, he had been compelled to stay on pri-
vately in A-tun-tsi till some of his debts were paid off, and
he had to sell a good deal of his property to accomplish
this.   Moreover the poor man was in a quandary owing
to his possessing two wives, a pretty Tibetan girl with no

money, whom he liked, and an ugly Chinese woman with plenty, whom he disliked. It required the utmost finesse to obtain aid from his lawful wife under these circumstances, for she was furious with the temporary Tibetan wife who had usurped her position, and made no secret of the fact.

However the liabilities were eventually met or evaded, and Hsia-fu was accorded a great send-off by Chao and the merchants, most of whom he had probably swindled. A table was spread just beyond the village, all the soldiers were drawn up in a line, and Hsia-fu feasted with the official and chief merchants who had thus shown him honour. Pretty speeches were then made by everybody, there was a shrill fanfare on the brazen trumpets, and Chao, preceded by his soldiers, rode back to his yamen, while Hsia-fu, having taken a tender farewell of everybody, set out on his journey.

On the 18th we started for Pang-tsi-la on the Yang-tze, a three days' journey. I had long wished to do some collecting on the pass and again compare the floras on the two sides of the watershed. In order to be as close to the pass as possible we went on till it was quite dark, and finally pitched camp on a grassy slope at an altitude of about 14,000 feet.

Leaving camp early on the following morning and taking with me only my Tibetan interpreter, we climbed to the pass (15,800 ft.) and struck up the ridge towards Pei-ma-shan, collecting seeds as we went.

About 1000 feet above the pass, that is at an altitude of nearly 17,000 feet, the screes began, and at this stage the alpine grass-land dwindled to an open formation which nevertheless included a wide range of species such as *Primula dryadifolia, Meconopsis rudis*, a *Scirpus*, an *Allium*, a Crucifer (*Cochlearia scapiflora*), and several cushion plants including a Potentilla and three species of Caryophyllaceae.

I have remarked elsewhere on the fact that most of the high alpines in this region have small seeds, doubtless to aid in wind distribution. But a second reason naturally suggests itself, namely the short time at their disposal for ripening, which would militate against the chance of their

receiving any great store of reserve food during the vegetative season.

It must be remembered that at 16,000–17,000 feet,
even on the comparatively dry Mekong-Yang-tze watershed, fine bright days are rare, while usually there are
damp swirling mists blowing up from the warm valleys
even when it is not actually raining. Under these conditions, insect visitors—whether butterflies, bees, or small
flies (and I have seen the last two at 17,000 ft.)—which are
already few compared with the number of flowers, are still
further restricted in numbers, if not altogether wanting, and
it is obvious that the flowers must therefore remain attrac-
tive for a longer period than if the conditions were more
favourable to pollination. This means that less time can
be allotted to the ripening of the fruit. Moreover a large
supply of endosperm would be of little value to a seed cir-
cumstanced as these are. Their best policy is to get ahead
as fast as they can immediately the snow melts, and make
the most of their ephemeral existence. At the low temperature at which these seeds must necessarily germinate,
the conversion and absorption of an oily or starchy endosperm would be a tedious business, and the simplest thing
to do seems to be to send out a root and leaf as quickly
as possible.

At all events it is significant that the majority of plants
at these altitudes are annuals or perhaps biennials, so that
their existence from the germination of the seed to the
ripening of the fruit is not prolonged much beyond four
months; nor is it likely, considering the rigorous conditions
under which they live, that they could absorb the materials
for, and manufacture a supply of, reserve material. All they
get they require for immediate use.

From the screes I finally tried to climb to the very
summit of the peak dominating this valley, a rocky pyramid from which a good view of the snow-clad Pei-ma-shan
group might have been obtained. But some hundreds of
feet higher, when I was a short distance from the top,
a stinging snow-storm swept down, blotting out the view
and so numbing me that I thought it best to abandon the
attempt, especially as the rock was rotten and some of the
traverses very nasty. I therefore descended to where my

guide was waiting for me below, and we headed back to camp.

At the very highest point to which I climbed there were still flowering plants in the shelter of the rocks, such as *Meconopsis speciosa*, cushion plants, and so on, but we may, I think, regard 18,000 feet as the extreme limit of flowering plants on this range. Above that the soil is probably too cold for the roots to function, since there is no reason why a plant should not obtain adequate protection from the wind at altitudes well above that limit.

Early on the following morning we struck camp and set out for Tung-chu-ling, the weather being fine and bright. After crossing the pass the great buttress of Pei-ma-shan was occasionally visible to the south, though partially buried in cloud, and immediately above us to the north stretched a barren ridge of limestone crowned by picturesque towers and pinnacles. No better proof of the difference in rainfall on the two divides could be adduced than by comparing the pyramidal peaks of K'a-gur-pu with their beautifully curved outlines due to water erosion, and the wall-sided, flat-topped buttress of Pei-ma-shan, plainly the work of dry denuding agents, particularly those associated with rapid oscillations of temperature and great extremes.

The snow clung to the precipices of Pei-ma-shan in patches and streaks, following particularly the lines of stratification, which were thus clearly defined; and the bottle-nosed glaciers, though several miles distant, were plainly seen to be retreating, their snouts being some little way from the terminal moraines.

Our route lay across a sort of plateau country, for after crossing the pass, instead of descending into the main valley, we kept on across the heads of several other valleys, ascending and descending steeply. Masses of a small brilliant blue gentian (*G. aprica*) covered the limestone rock, which here presented a remarkable vesicular structure. Very conspicuous also at 14,000–15,000 feet was the curious glaucous-leaved *Spiraea laevigata*, which predominated in the shrub flora. It forms loose-growing bushes about six feet high, but unfortunately it was not now in flower.

A terrific and sudden storm of rain coming on at mid-day, we sought the shelter of some tents occupied by yak

W. T.

11

herdsmen, and had lunch.　We soon reached the summit of the last spur, and saw below us to the east the blue mountains of Ssu-chuan, and the deep valley far beneath in which flowed the Yang-tze.　Fine forests of larch and fir filled the higher valleys.

It was dusk when we reached the first huts; away down below us the setting sun glistened for a moment on the golden spire and pale walls of Tung-chu-ling monastery, and shortly afterwards we were enveloped in darkness and continued thus for two or three hours by a most dangerous path above the torrent.　Every moment I expected the pony to slip and go over, but happily I never saw the more appalling places till we were there, when the only thing to do was to sit still and pray.　Finally I had to get off and lead my pony, sagacious little chap though he was.　At last my guide lost the road altogether, and as he stood there on the mountain side arguing with the porters I went ahead myself, for I have a way of finding the path on the darkest night by instinct.　I managed to hit it off, and arrived at the scattered village of Tung-chu-ling, where we found comfortable quarters in a large if unpretentious house.　It was warm down here, and scarcely waiting for supper I turned in and slept like a rock.

I awoke in a new world.　Fields of waving millet and maize, tall runner beans, immense vegetable marrows, and huge nodding sunflowers greeted the eye, but the big monastery perched on the summit of the hill above us, though scarcely visible from here, was the chief feature of the place.

Sending the porters on ahead to Pang-tsi-la I went with my guide to visit the monastery, a large rambling old place surrounded by a high wall.　The lamas received me with evident surprise, but were outwardly friendly as I walked round the courtyards, explored the cells, and entered the main temple where a service was being held in the usual noisy fashion.　There was little of interest in the monastery itself, however, except a large and crude wall-painting of one of the big monasteries at Lhasa, done in the most quaint perspective, or rather lack of it; and I finished my inspection by climbing out on to the roof of the temple, which was covered with wooden slats held down by rocks just like the Lutzu huts on the Salween. Crowning the whole were the usual copper or brass ornaments, like mock chimneys or cowls, decorated with various symbols of Lamaism, and in the middle a golden cupola which, when the sun shines on it, is visible from a great distance. There are some 300 priests belonging to the monastery which, owing to its isolated position, looks far more imposing when viewed from either up or down the valley, being but a scurvy place when seen at close quarters.

Higher up the hill side is a smaller women's monastery, if I may use the term, women priests being not uncommon in this part of the country. However, though curious, I thought it might be improper to pry too closely into the matter, and did not suggest visiting it.

The descent to the Yang-tze was through very arid country until the terraced and irrigated alluvial fan sloping from the valley mouth was reached. Here were crops of cotton, tobacco, hemp, and fields of barley, pomegranate trees laden with glowing fruit, peach trees, groves of orange bushes, and walnut trees, amongst which were scattered large white houses.

The women do not in the least resemble the Tibetans of the Mekong valley, being short of stature with remarkably small features—*petite* is perhaps the word which describes them best. They wear pleated skirts like those in vogue amongst the Mosos, and I am inclined to think that they are of Moso rather than of Tibetan origin, though they speak the latter tongue. The whole of the mountainous region east of the plateau is inhabited by more or less isolated Tibetan tribes, differing from the real semi-nomadic people of eastern Tibet and from one another, but evidently Tibetans in the broadest sense of that term, and the ancient and once powerful Moso kingdom which had its capital at Lichiang-fu may have grown out of the most successful of these tribes. All the way up the Mekong valley from Tsu-kou to Y'a-k'a-lo we meet with strange Tibetan tribes, but the mixture along the accessible roads of these border regions is now so great, comprising both hideous negritoids and really beautiful girls, that any attempt to disentangle the elements seems at first sight hopeless.

Between Pang-tsi-la and Tung-chu-ling are several tall mud watch-towers standing up above the house roofs, another monument to Chinese activity along the trade routes, though it must be confessed that, now at any rate, this road is little used by any but the lama caravans going to Lhasa, and the Tibetans have it all their own way, in spite of the fact that since the 1905 rebellion there have been Chinese garrisons at Pang-tsi-la and Chung-tien. This road offers the quickest route between Tali and Batang, but except when in large numbers, the Chinese always prefer to go by Wei-hsi-ting, and I do not remember to have met a single Chinese caravan between A-tun-tsi and Pang-tsi-la. The Yang-tze here is narrower than below Batang, and not interrupted by rapids to any great extent, so that except in times of unusual or sudden floods it can be crossed by the ferry. Barren stony platforms—sometimes cultivated if water is available, as below Pang-tsi-la itself—extend from the mountains to the river, above which they end in sheer gravel cliffs. It had been my intention to continue up the Yang-tze to Mo-ting and return to A-tun-tsi over the Rün-tsi-la, but learning that this meant a journey of five or six days in the arid region, I abandoned the idea, and decided to start back next morning by the same route, spending another day collecting on Pei-ma-shan.

Next morning I could not secure any porters, and whereas Gan-ton would have pressed half-a-dozen into service in a very short time, hour after hour passed and still my juvenile interpreter had not collected the requisite number. Finally starting at one o'clock, it was dark long before we reached Tung-chu-ling and we had the pleasure of another night march. Eventually everything arrived safely except my bedding, and a search party was sent out with torches. We found one of the porters had quietly given up the struggle in the darkness and sat down.

On the following day, September 23, we again started late, and ascending the steep valley in leisurely fashion, made only a short stage, camping at the upper limits of the fir forest, and on the third day after leaving Pang-tsi-la we camped just below the pass at an altitude of 15,000 feet, the wind here being bitterly cold.

As I was riding across the plateau some distance behind

the men, two Tibetans advanced, one of whom, with never a smile or a bow, caught my pony by the bridle and thrust into my face a wine bottle, requesting me, in violent language as it seemed, to drink. Raw spirit however is not at all to my liking, and when I politely refused, showing none of those qualms I was really feeling, he became still more aggressive, fumbling inside his voluminous cloak for something—a sword I began to suspect! Anyway I thought it best to humour him and sipped at the ice cold spirit, which made me gasp and choke. Finally I handed back the bottle, refusing to drink more, and the stranger, letting go of my bridle, departed as he had come, shouting rather than saying things over his shoulder as he went.

Above Tung-chu-ling pine-trees were scattered amongst the limestone rocks as at Tsu-kou, and as always where pines occurred, flocks of green parrots darted screeching from tree to tree. They come after the red berries of a semi-parasitic shrub belonging I believe to the order Santalaceae, or perhaps a *Viscum*, and I often watched them rubbing their beaks on the branches to get rid of the viscid seeds, even pulling them out of each other's mouths, a most droll performance. Under some of the trees one could frequently pick up dozens of these seeds and half-devoured berries.

It had been my intention to start at the first sign of day and climb Pei-ma-shan before the clouds rose out of the valleys; for this reason I had pitched camp near the pass, though our exposed position made it miserably uncomfortable. However in the night I had a violent attack of sickness, and being scarcely able to touch any breakfast, all ideas of mountaineering had to be given up. I therefore gave the order to return to A-tun-tsi, and though racked with horrid pains so that I found some difficulty in clinging to my pony, we reached our base camp in the evening and I soon felt better. On the way down I found masses of a most brilliant blue trumpet-shaped gentian (*G. ornata*), a typical limestone plant, just coming into bloom at the end of September.

The first news I heard on my arrival was that my landlord's small daughter, a merry little girl of twelve, had fallen off the roof and been killed, a disaster which so overcame the father that he was generally to be found drunk in the evenings from that day onward.

Meanwhile we settled down to routine work once more, relieved by one or two minor festivals, visits from various merchant friends and soldiers, people asking for medicine, and Tibetan dancers. These latter were dressed in the most brilliant robes of green, blue, purple, and scarlet, with aprons of jingling bells fastened round their waists, and the women always carried round their necks large amulet boxes of silver set with coral and turquoise. They sang in a harsh voice while the men footed it gaily to the shivering of the bells and the beat of drums. Sometimes a sorcerer appeared, anxious to tell everybody's fortune and cast out devils; there must have been any number of the latter lurking in the village for he was always in great demand. Then the village barber, who had once or twice scraped my face, came with a hog bite six inches long and a week old in his arm. He had killed a pig for bacon, but the redoubtable beast before dying had succeeded in leaving his trade-mark on the barber's arm.

Knowing on what these scavengers feed I was surprised that the wound, which had been liberally smeared with butter, was not more loathsome than it was, and giving the patient some permanganate to wash it with, and telling him to poultice it with *tsamba*, I dismissed him, having now no time to spare over bizarre medical experiments. The wound healed in a week, but whether because of the butter or the poultice I cannot say.

So the last days of September slipped by and autumn crept down into the high valleys.

CHAPTER XIII

OVER THE RÜN-TSI-LA: A THIRD JOURNEY TO THE YANG-TZE

THE cultivated slopes just above the village were now bare, and the place looked dismal enough, for the corn was already reaped and the Tibetans could be heard singing to the rhythmical rise and fall of the flails on the neighbouring house-roofs. The leaves were rapidly turning yellow and falling from the trees, the flowers dying, and winter was creeping over the mountains, but the weather was not yet settled and showers were frequent. Sometimes the entire valley was buried in cloud so that it was impossible to see the length of the village, and then a persistent drizzle filled the air. In September we had seventeen rainy days, on three of which it rained all day; and between October 1st and 17th we had thirteen rainy days, on six of which we never saw the sun at all, as rain fell almost continuously.

There are two rainy seasons in A-tun-tsi, the first in early spring, usually before May, the second in late summer, usually ending some time in September, and this year the second rainy season had been abnormally prolonged. It is this late summer rain which accounts for the rich autumn flora of the Mekong-Yang-tze divide, saxifrages and gentians being found in full bloom at altitudes where deep snow is already lying on the Mekong-Salween divide. The flora of the latter ridge is essentially a summer one, attaining its maximum development in June and July, for the rains break there before the end of June. Twenty miles to the east as the crow flies, the flora reaches its maximum development two months later.

Time now hung a little heavily on my hands. The

flowers were over, Kin having wound up the list with a delightful gentian (*G. Georgii*), and only the seeds remained to be harvested. I climbed the mountains as usual, but not so eagerly as of yore. Sometimes I went out early in the morning with my gun, but though I occasionally saw the common pheasant, I rarely shot anything larger than a pigeon. The more interesting forms of game, such as snow pheasant and other kinds, grouse, partridges and so on, I invariably came across when I had no gun with me. M. Perronne however, who had with him an excellent pointer, was more successful and shot a good many pheasants, several of which he sent round to me from time to time, and I found them a very pleasant change of diet.

Then came wild rumours from Ssu-chuan. Runners had arrived from Batang with news of a big rising in that province, but previous to this there had been several unauthorised versions of the story going about in the village. Kin told me that a lot of Chinese soldiers had deserted and taken to the mountains; later I was told that the Tibetans and Chinese were fighting again; and now on October 12 we had the first news of the great revolution in China, though even then we did not realise what it meant. The merchants told us that the reform party had risen in Ssu-chuan, that 3000 of them had sacked Ya-chou-fu, that the Europeans were all leaving China, and finally that Chung-king, the great river port on the Upper Yang-tze, had been destroyed.

There was of course likely to be some foundation of truth in these vague reports, but the extent and the significance of the reform movement were quite misunderstood in the poor little village of A-tun-tsi, which, being off the line of the posts and telegraph, remains in complete ignorance of what is going on beyond, unless of course the interests of the community are directly threatened. Moreover the official being away, we could get no information from that source either, and meanwhile I was chafing for action. I had harvested most of my seeds, and was restless to be up in the mountains again, partly in the hope of securing seeds of plants I had seen on my travels which did not, so far as I knew, grow in the neighbourhood of A-tun-tsi, and partly in order to secure seeds of some of the highest alpine plants whether I had seen them in flower or not, since they were likely to be of interest from a botanical point of view.

It was now or never, if I wished to cross the Mekong-Yang-tze divide by the pass which leads to Mo-ting; for Chao and the senior military officer—a dandy from Yunnan-fu who wore a smart Japanese uniform with a long sword and white kid gloves—had gone on a tour, and there was no one in A-tun-tsi with sufficient authority to prevent me. The man from whom I hoped to obtain permission was the junior military officer, a friend of mine who often came in for a chat, to look at my automatic pistol and other marvels, or to beg some gun oil. He would not be likely to refuse my request. True, Chao himself might have given me permission to go, for he had left me to my own devices recently; but he had several times complained bitterly that I always wanted to go off the main roads to the most obscure places, and the path to Mo-ting was used only by a few Tibetan caravans going into the very heart of the Mantze mountains.

I grew restless as on moonlight nights I heard the steady thump, thump of the flails on the house-roofs near me, the droning songs of the Tibetan girls, a shout of laughter or snatches of conversation; restless as I worked in my room drying and packing seeds; still more restless as I climbed the now familiar mountains in the chill October drizzle swept in my face by fierce gusts of wind. The high mountains which shut in the village were already white with snow again, and so cold was it that I always had a big *ho-p'an* of red-hot charcoal in my room.

By October 17 everything was ready, and on the 18th we started for Mo-ting, Kin remaining behind to collect seeds and attend to those that were drying. For interpreter I had a local Tibetan who proved very satisfactory, and subsequently requested me to employ him permanently should I return to A-tun-tsi.

Our route lay up the valley I had already ascended several times, but instead of turning off into any of the hanging valleys which opened into it, we kept on up the main valley. Accompanied by one of the men, I ascended slowly to within sight of the pass, and stopped by a small lake which partially occupied a deep depression, intending to camp here for the night. There was dwarf rhododendron for firewood (our altitude was about 16,500 feet), fresh water in abundance, and a well-sheltered grassy hollow in which to erect the tents out of the wind and away from the patches of snow which lay here and there. But the porters were a long way behind, and when after half-an-hour's waiting I sent my man back to seek them, it began to snow heavily; it was already late in the afternoon, and the short autumn day was drawing to a close. In front rose the rugged snow-clad crest of the watershed, and on either hand the bare screes stretched up to the limestone towers above. The lake was quite small, but what particularly drew my attention to it was the outflow, for a solid wall of rock which cut it off from the valley forty or fifty feet below had been sawn, as it were, clean through for perhaps fifteen feet to the level of the water, which issued between two stone portals a few feet apart, and fell over boulders in a cascade for the remainder of the distance. The water in the basin was a beautiful deep green in colour and reflected the snow mountains above in the most glorious manner; it was half frozen over however, and the narrow exit was choked on either side with icicles. Whether the water really had carved out this channel through the solid rock which hemmed in the basin, I had no means of judging, but I do not think it impossible. However in this case there was no room for doubt that the lake occupied a rock basin; moreover, coming up the valley we had already passed two other pools, finding one at each level as we ascended the long steps of the valley.

After climbing round it, finding numerous rock plants and scaring up a giant mouse-hare, I returned to the shelter of the hollow, and amused myself by collecting a big pile of rhododendron for firewood; but the snow-storm getting worse, I could stand the cold no longer, and started down in search of the missing men. It was half-an-hour before I fell in with them and as it was impossible to reach my camping ground before dark, the only thing to do was to descend to the first lake and camp by it, which we did, pitching the tents above the tree limit at an altitude of about 15,000 feet.

It was a glorious night, the sky powdered with myriads of stars, but the cold was intense. The dark figures of the men curled up round the leaping fire looked ghostly in that lonely mountain valley. I could just hear the murmur of the torrent, and occasionally the tinkle of the ponies' bells as they hobbled about looking for grass; otherwise the night was extraordinarily still, and I was just dropping off to sleep when I heard the Tibetans running about and shouting; apparently the animals had strayed or another party had joined us. Then the canvas of the tent began to buckle noisily as it froze, and taking it all round I slept badly, though with heaps of woollen clothes on I kept warm beneath my five blankets.

Next morning in brilliant weather we started early for the pass. The frost had been severe, the path was like iron, and there was much more ice about. Scattered along the stream-side were dwarf willow and occasional clumps of a large-leafed *Senecio* which ascends right to the furthest limit of plants.

It is remarkable that while trees extend to higher altitudes on the valley sides than on the valley floor, shrubs on the other hand extend to higher altitudes on the valley floor than on the valley sides. The reason is doubtless to be found in the fact that, towards the summit of the watershed, wind and water are both concentrated in the valley bottom, the former being inimical to trees, while at higher elevations the latter is essential to the shrub vegetation. Thus, towards their limit, trees are driven to occupy sheltered places on the mountain slopes, while towards the limit of shrubs, the dwarf vegetation which is able to withstand the wind is driven to occupy the valley floor, rhododendron giving place to willow.

Just above the lake I had found on the previous day we came into the snow, which grew rapidly deeper, but happily the party which had joined us on the previous night and gone on ahead had stamped a trail, otherwise we should have found it desperately hard work.

The last few hundred feet were up a zigzag path, the precipitous rock wall rising abruptly above us to a small notch in the ridge, which was the Rün-tsi-la. The snow here was very deep, and nowhere on the exposed rock was there any trace of vegetation, which seemed to end several hundred feet below the summit; consequently I think the Rün-tsi-la can hardly be less than 18,000 feet. But though I was very much out of breath and had to stop every few yards, I felt none of the effects of mountain sickness on this, the highest pass I crossed.

The last vegetation consisted of patches of *Potentilla*, dwarf *Meconopsis* (*M. rudis*), dwarf *Primula*, *Scirpus*, a few grasses, a gentian actually in flower (*G. heptaphylla*), a *Senecio* at the very head of the trickling stream, and several scree Compositae and Umbelliferae.

Under the combined action of sun and wind the snow had assumed some picturesque forms, and in the shade of the rocks small ice-pillars stood up from the ground, evidently formed in the following manner: the snow on the boulders melted in the sunshine, trickled slowly over the rock and froze again as soon as it got into the shade, an icicle being thus formed in the usual way. Eventually it reached the snow, and then the end in contact with the rock thawed as the sun got round so that the pendent icicle became an erect ice-pillar. At the summit, the snow on the south-facing rocks stood out in horizontal fern-like plates and crystals of great beauty, often several inches long. The manner of their formation I could not make out, but it was probably connected with the almost simultaneous melting of the snow in the warm sunshine, and its freezing and recrystallising in the bitter wind.

The descent through deep soft snow was bad, and spills were frequent. At last we got down to a small lake in the stony valley and halted for lunch beside a party of Tibetans on their way up. The men now informed me that there was a choice of routes, either down the valley in front of us to the village of Ngoug-chi and so up the Yang-tze, a day's march to Mo-ting; or over a second pass and down the next valley. However I was chiefly concerned with the mountain flora, so I decided to go straight on.

This second pass, the Chnu-ma-la, seemed scarcely as high as the Rün-tsi-la, though the less amount of snow on the south face was easily accounted for by its much more exposed position; while on the north slope the snow was almost as deep as on the Rün-tsi-la. From the summit we

Plate XXXI

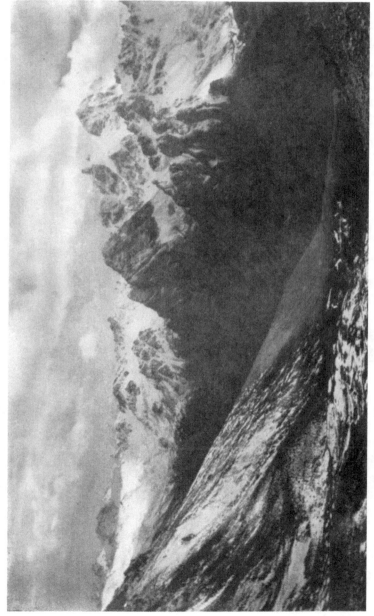

The Mekong-Yang-tze Divide, looking south from the Chnu-ma-la

obtained a grand view of a big snow mountain on the main watershed due south, but its name I could not ascertain. I had seen this same snow peak from the mountain to the west of A-tun-tsi, and its position is almost due east of that village. With the exception of Pei-ma-shan, some miles to the south, it is I believe the only snow mountain between the Mekong and the Yang-tze, south of Batang ; and considering the abnormal height of the snow line, it cannot be less than 21,000 feet, and may be considerably more. To the north and east the view was very different, for here was spread out at our feet a sea of blue mountains with dark lines of forest marking where the deep troughs separated crest from crest of the mountains of Ssu-chuan.

Our route still lay northwards across the heads of several small valleys, but we had left the bare stony slopes behind and were amongst the shrub vegetation again. Not far to the east rose a great cliff of a bright red colour, glowing in the evening sunlight, and forming a very conspicuous landmark. Fragments of this rock scattered about showed that it was a sandstone or arkose.

Camp was pitched in the next valley at an altitude of about 15,000 feet, just on the edge of the larch and fir forests. A heavy snow-storm assailed us as we were putting up the tents, but it soon passed and a brilliant night followed. I found it difficult to sleep at these altitudes and was restless in consequence, though I managed to keep fairly warm. Next morning frost glistened on everything, and the larches, sere and yellow, looked very pretty in the sunshine. Two curiously-shaped limestone peaks, separated by a low saddle, were conspicuous objects to the west, and beyond them was a higher peak covered with snow.

At midday a long descent brought us at last to the first signs of habitation, and here we found a hot spring. Continuing through forests and gorges, the deciduous-leaved trees finally gave place to *Pinus* and scrub oak, but before long trees ceased altogether, and we entered the arid regions once more. In the middle forest, at an altitude of about 10,000 feet, I found a red-flowered saxifrage growing on the rocks in deep shade.

The Mo-ting Tussu, a tall, strongly-built Tibetan with a wrinkled pleasant face, came out to meet us and led me to his house. It was delightfully warm now after the fierce cold on the passes, and so pure was the air that the Milky Way looked like a silver band stretched across the sky.

Mo-ting, with a population of forty or fifty families and a tiny lamasery of no consequence, is built on one of the wedge-shaped spurs blocked out by the mountain torrents and cut deeply into below, so that it presents a comparatively gentle slope from the mountains to the apex of the wedge, and steep precipices on either side. It is a thousand feet or more above the Yang-tze which is a mile distant, the aneroid reading 21·45 in. at my house and 23·07 in. in the river bed, a difference of 1·62 in.

The river is narrower than at Batang, resembling in some respects the Salween or Mekong in the arid regions, so that some of the differences pointed out in Chapter X do not apply here. It flows between tremendously high and steep cliffs which are scarred and torn as though by furious rains or sudden torrents, and is quite unnavigable. Caravans travelling north-east to Litang or Tatsien-lu no doubt cross lower down, but even then probably by ferry, as at Pang-tsi-la, and not by a rope bridge.

The Mo-ting slope is terraced, the terraces being irrigated by means of a mountain torrent diverted into side-channels. Quantities of hemp are grown, this plant apparently doing very well on the granite, and growing to a height of ten feet; there are also crops of millet, wheat, barley, and buckwheat, while walnut trees, pomegranates, oranges, and a peculiar little persimmon make these villages in the arid regions conspicuous from afar. The prickly pear was grown in places.

In the afternoon we started for Yie-rü-gong, a village about ten miles up the Yang-tze, by a path which kept between one and two thousand feet above the river, but so straight-sided was the gorge, that though we could not hear we could generally see it. On the other side the villages perched far up above the river were occasionally visible, but there was no room for any habitation below, and the valley is very sparsely populated.

As darkness came on the road became worse and worse; here we ascended by roughly-laid zigzag steps beneath the shadow of overhanging cliffs whose summits were lost in the gloom, there the path plunged steeply down towards a bridge of tree trunks spanning a gully. In some places it was indeed horribly dangerous, consisting only of bare slabs of rock tilted downwards, so that a slip would have shot one over the precipice, and on one such occasion I crawled across on my hands and knees. I do not know which were the more terrifying in the darkness, the precipices above or those below. Needless to say I did not ride my pony after dark, and it was ten o'clock before we reached Yie-rü-gong.

Our night's lodging was paid for in brick tea, which the Tibetans in the mountain villages off the main roads always prefer to silver, since it is immediately useful, whereas the latter is not ; and I had by this time learnt to carry brick tea for that purpose. Silver is certainly not valued by the Tibetans for its own sake.

About 3 a.m. I awoke to see what looked like a moonbeam across my bed, though I had seen the moon set quite early, and stepping out on to the roof I saw a planet of wonderful brilliance, presumably Venus, rising in the east. So bright was the light that I could have seen to read large print by it.

After leaving Yie-rü-gong we continued up the Yang-tze for another half-day's march and then struck up a ravine to the west in order to recross the mountains and reach Tsa-lei on the Batang road. An alternative route was to continue up the Yang-tze to the mouth of the Garthok river, and so to the Batang road where it joins that stream. This indeed was what the men wished to do, as they maintained that the pass to Tsa-lei was too difficult. However by paying them their own price (certainly not an exorbitant one) we persuaded them to try it, for there was nothing to be gained by continuing up the arid Yang-tze valley.

It was October 22. As I rode along I gradually became conscious of a peculiar appearance in the sky to the north, which had assumed a deep violet-blue tint. At this time I wished to take some photographs, but the waning light made me hesitate; yet when I glanced at the sun I was surprised to see it apparently shining as brightly as ever, a thing I could not understand, for I could have sworn that it was obscured by clouds. There were no clouds however, except far away in the south, where it looked like late evening, so subdued was the light. It grew darker and darker, and at last it dawned on me quite suddenly that this was an eclipse of the sun, a phenomenon I had never before witnessed.

The eclipse was almost complete, the sun showing as a red disc surrounded by a narrow silver rim. Seen through a film of cloud the dark shadow of the obstruction was visible passing over from above downwards, and seen through snow glasses it appeared quite black. The eclipse began about 9 a.m. and lasted for two hours or so, our position then being roughly 28° 50″ N., 99° 15″ E. The Tibetans had taken no notice of the growing dimness, but when they saw me staring up at the sun through blue glasses they began talking a good deal amongst themselves, though they were far from exhibiting signs of consternation and scarcely evinced surprise.

Dropping down to the bottom of a deep ravine we began the ascent, the torrent here taking a series of immense leaps down a colossal granite stairway, hemmed in by cyclopean walls of the same rock. At first we scrambled up vast screes of rough blocks, through dense shrub vegetation similar to that in the arid valley below, but in the shade of these cliffs growing far more luxuriantly, with festoons of Clematis and twining *Polygonum* hanging from every bush, and a rank herbaceous undergrowth all around. Then came scattered maples and willows, oaks, poplars, birches, elms, and other trees, with pines clinging desperately to the precipices above. Just beyond the last cataract lay two or three immense blocks of granite fallen from the crumbling cliffs, one of which I estimated to contain at least 20,000 cubic feet. Riding between them was a dangerous game, for in some places the pony jammed, and I almost had a leg crushed.

Beyond this point the valley began to broaden out, and quiet reaches followed, where there were groves of poplar and walnut trees. As already pointed out the mountain torrents in the arid regions reverse the normal sequence of valley structure. Instead of gorges at the source of a river opening out into a valley which grows wider and wider till at length a flood plain is formed, we find a wide valley in the mountains gradually contracting till finally the stream cuts its way through a narrow ravine shut in by perpendicular cliffs; and this irrespective of the kind of rock met with. The reversed valley is however purely a climatic effect, at least in this region; for heavy rains occur in the mountains at the valley head and the torrent flows down into the rainless arid region where nothing but gorges can be formed.

We camped on a grassy flat amidst poplars and firs at an altitude of about 12,000 feet and awoke to find it snowing.

Once *en route*, riding soon became intolerable, for my feet and hands were nearly frozen, so I clambered up through the forest on foot. When we stopped for lunch under some junipers near the tree-limit the snow was whirling down thicker than ever; nothing was visible but grey clouds, trees laden with snow, and a world of falling flakes. Through the heavy mists everything loomed white and indistinct, and once out of the forest we got the full benefit of the wind. Granite had given place to the usual limestone capping this range, but of the fine scenery we were now coming into I could see little. The valley head was blocked by a rounded hummock of rock over which the stream poured in several small cascades, but this obstacle we outflanked, reaching an open plateau-like valley covered with dwarf rhododendron, but now buried under deep snow. We could not see more than a hundred yards in any direction, but presently we entered a narrow stony gulley with tremendous scree-slopes rising steeply on either hand to the splintered limestone towers which had given birth to them. It was a most desolate scene and the going was very bad, for the gulley being blocked by large boulders, it was necessary to traverse the scree itself.

Finally we scrambled up a wall of rock to the actual pass, where our troubles really began, for we had turned to the north. It was difficult to arrive at an approximation of the altitude from a comparison of the vegetation on the two sides, on account of the snow, but evidently we were somewhere near the limit of plants on both north and south slopes, so that 17,000 feet is probably a near estimate for the Ā-löng-la, which is considerably higher than the Tsa-lei-la, marked 15,800 feet on Major Davies's map.

This spur separates the water flowing directly into the Yang-tze near Yie-rü-gong from that flowing to the Garthok river, which enters the Yang-tze further north ; hence the A-löng-la does not lie on the main watershed. A second pass was however visible scarcely a stone's throw to the south, and this I think must be on the main watershed.

Across the path the snow was much deeper, and very soft ; men and ponies slipped on the loose scree, and we had to go dead slow. But it was the wind whistling over the passes and dashing the snow in our faces which made me call myself a fool for coming and vow I would never do such a thing again. Sensation almost left my feet after a time, but I continued to ride my pony, having no desire to slither about on these treacherous screes and roll in the snow as the men frequently did, though more than once the pony looked like turning a somersault.

I noticed a lot of *Meconopsis speciosa* lower down, but somehow the wind rattling amongst the dead haulms gave me a momentary distaste for botany. Below the screes was a fair-sized frozen tarn ; the shrieking gusts of wind sent the dry granular snow humming over its surface, and to listen to this dismal sound, to see the pale mountains looming indistinctly through the mist, and the white caravan picking its way carefully down the valley, was to think of Sven Hedin's description of the Chang Tang. I was glad I had only a few hours of it instead of a few weeks!

Below the lake the valley began to open out and dwarf shrubs appeared ; we turned more to the west, and the snow ceased abruptly. Just then the sun shone out momentarily and gleamed on the white mountains, blue sky appeared above, and behind us we caught a glimpse of scudding clouds. Far away down the valley in the west we saw over the dark forest a heavy bank of slate-blue cloud, but the setting sun nevertheless painted the light clouds above us, and everything looked well. Slowly the snow-drenched caravan of seven men, one woman, two ponies and a donkey, now spread over a quarter of a mile, struggled down the valley.

Darkness set in as we entered the forest, and some of the men wanted to camp, but we would not hear of it ; Tsa-lei could not be very far away, for we could see the main valley ahead. Under the trees however it was soon pitch dark, so that we had to proceed with the utmost caution, the path skirting a deep ravine. In order to keep awake and at the same time relieve the porters, I went in front leading my pony, and on one occasion having lost the path and finding myself on the sloping face of the cliff, I had to crawl for safety on my hands and knees. In some places it was impossible to see five yards ahead, and we were lucky perhaps to get down without accident or the loss of an animal.

At last we emerged from the forest on to level ground to see the sky blazing with stars again, and then we found ourselves amongst some black shapes grazing in a meadow. At the same moment came that most familiar of all sounds on approaching a Tibetan village at night—the deep baying of dogs. About 9 p.m. with wild war-whoops a weary but elated party crawled into the little village of Tsa-lei, and exhausted after our twelve hours climbing we found shelter in the hut where we had slept the night four months previously, on the road to Batang.

There was the same difficulty and delay in getting off next morning that we had experienced on the previous occasion, and my advice to travellers is to give Tsa-lei a wide berth. I paid off the Mo-ting men, and to their great delight gave them something extra as they had done so well on the previous day, but I could not prevail upon them to go with me for another stage. Moreover the Tsa-lei trade-union of porters, as it were, strongly objected to Ssu-chuan porters carrying my loads in Yunnan, though they were in no hurry to do it themselves. It was ten o'clock before we made a start, and some of the loads did not leave the village till mid-day.

The sea-buckthorn trees lining the stream were now a mass of orange berries and the Tsa-lei-la was under fresh snow, though nowhere was it very deep; probably it would melt on the south face in the course of the next day or two. From the summit we looked southwards over a magnificent panorama of snow-clad mountains, amongst which the pyramid of K'a-gur-pu was a conspicuous object.

We camped for the night at the edge of the forest once more and next evening, October 25, we reached A-dong.

12—2

The first news I heard on my arrival was that the missionaries had left Batang! This was a bolt from the blue with a vengeance! Two Europeans were reported to have arrived in A-tun-tsi on the previous day, and two more, whom I learnt later were American missionaries, were even now sleeping in A-dong, having arrived the same day as myself. However I decided not to disturb them as they were likely to be tired after their journey, and certainly I was. I therefore went to bed, expecting to see them early on the following morning, but I was so late that they were up and away long before me.

On the way to A-tun-tsi I fell in with a soldier who had been sent by the official to seek me and escort me back, on account of the trouble in Ssu-chuan, a kindly act on his part.

We reached A-tun-tsi at mid-day on October 26, after a journey lasting only nine days, though it seemed much longer, no doubt on account of the cold. On the whole I had been fairly successful with my work, securing seeds of several new plants; but alas! most of the photographs taken on this journey, together with all of those taken on the journey to Pang-tsi-la, were subsequently lost.

CHAPTER XIV

A WINTER JOURNEY AMONGST THE LUTZU

On my arrival in A-tun-tsi I went straight down to the other end of the village to see M. Perronne and learn the news, but just outside I was hailed by Mr Edgar, who with his family was installed in the Tussu's house below the village. Almost the first news I heard from him was that Captain Bailey had got safely through to India. Mr Edgar told me that the Batang posts were disorganised and the telegraph line cut; all the missionaries had consequently left the place and were going down to the coast. There were ugly rumours concerning the fate of some English missionaries who were isolated somewhere in the Tibetan Marches and a party of French priests were also reported missing; but as a matter of fact all the Ssu-chuan missionaries eventually reached the coast in safety.

There were now no less than eleven Europeans in A-tun-tsi, comprising the English and American missionaries from Batang with their wives and families, M. Perronne, and myself. In the evening we held a council of war, several Chinamen prominent in the village attending, but on the one subject on which we desired information, namely the condition of Yunnan, not a word was forthcoming. It was suggested that we should leave in a body, as a measure of precaution, but this was impossible, since the village could not supply sufficient transport for more than one party to start at a time, and on the following day the American contingent left.

Meanwhile Kin and I were finishing up our work, for I wanted to leave on November 1 as originally planned. To put it off any longer might be fatal to our chances of recrossing the mountains to the Salween, for already the passes were reported under deep snow.

On October 28 the official returned, but if he knew anything, which was doubtful, he kept it discreetly to himself; and two days later Mr Edgar and his party started.

The day before I left an unfortunate thing happened. By some means or other it leaked out that I was bent on returning to the Salween valley, and in the afternoon the story was all round the village. It was the more annoying because I had purposely refrained from saying anything about it even to my men, thinking it best to say nothing and just go. Probably my attempt to secure the services of a Tibetan interpreter—in which, by the way, I failed—aroused suspicions and caused someone to jump to a fairly accurate conclusion as to my real intentions. For one does not need a Tibetan interpreter for a quiet journey down to Wei-hsi, which was my professed destination after leaving A-tun-tsi. Be this as it may, my friend Chao the local official heard the rumour and at once sent round a deputation forbidding such a journey. He would be very pleased, he said, to give me men and animals to see me safely down to Wei-hsi, but if I persisted in my plan of going to the Salween he would lay an embargo on all transport and thus give me checkmate.

In the afternoon I paid a farewell call on the official, who had treated me with kindness and consideration during my stay in A-tun-tsi, and I asked for permission to go to the Salween on my own responsibility, a suggestion which met with no response. The interview was then terminated by Chao who said politely, when bidding me good-bye, that he hoped I would come again next year.

On the following morning, November 1, the animals came round to my house with an escort of three soldiers, and before nine o'clock I had finally turned my back on the little village which had been home to me for six months. The only European now left was M. Perronne, who stayed boldly at his post, but he had grown too much accustomed to the place and its people to fear anything. I often wondered how he fared in those troublesome times, but towards the end of May I received a letter from him written in Rangoon, saying that he was all right.

As a parting gift I sent Chao a silver-topped bottle from my suit case, the only respectable article I had left to give away, and as he accepted it I thought it highly unlikely that he would interest himself further in my plans. He had protested, as he was in duty bound to do no doubt, and there the matter ended. However, for the time being I kept my own counsel.

The fine winter weather had at last set in, and the days were superb, but while at mid-day in the sunshine it was quite hot, no sooner had the sun sunk behind the lonely monastery, than it grew very cold, and the brilliant nights were bitter.

As we rode down the valley, it was interesting to observe how the leafless trees above A-tun-tsi were gradually replaced by gorgeous autumn tints and scarlet berries, giving place to green trees, and finally in the Mekong valley itself to autumn crops of buckwheat.

The two days' journey to Yang-tsa was without incident, and it then became necessary to decide on a plan of action. I therefore told Kin that we would cross the river at Yang-tsa, sending the rest of the caravan on to Tsu-kou, there to await our arrival, while he was to speak quietly to one of the soldiers who understood Tibetan, and bribe him to accompany us as interpreter. At first the soldier hesitated, saying that Chao would beat him when he got back, which was very likely true as it was well known that I had been forbidden to go. But when I pointed out that Chao, being a friend of mine, would only give him perhaps fifty blows, a merely nominal punishment to 'save his face,' he consented on condition that I made it worth his while, and my offer was at once accepted.

To one of the other soldiers I gave a small present of silver, telling him to take the rest of the caravan, which, with my pony Beauty, was in charge of Sung, to Tsu-kou; and next morning we split into two parties.

Had the soldiers, acting under orders, refused to countenance this arrangement, I should have collected all the slings in Yang-tsa and quietly slung the baggage over the river in the dead of night. Kin and I would then have crossed and started at daylight, leaving the rest of the caravan on the other side.

After watching the departure of the Tsu-kou party, I told our soldier to see the kit across the river, and find porters to carry it up to the village of Londre, whither we had descended from Doker-la in June; but owing to a misunderstanding nearly half of it was left behind, and he went off to the village, a good three hours' walk, leaving it and us to our fate.   Kin therefore stayed behind to look after the baggage and engage porters to carry it up, but it was ten o'clock at night before he arrived with everything, though in the meantime I had gone on to Londre and sent men back to his assistance.   It was not an auspicious beginning for I had no lunch and only milk for supper on account of this business.   Also, we were no sooner across the Mekong and south of Yang-tsa than it began to rain, and at this season rain in the valley meant snow on the mountains.

Next morning our soldier, who seemed very much afraid of the Tibetans, wasted four hours finding four porters willing to cross the pass which, though only 13,000 feet, was already under deep snow ; more we could not get, so something had to be sacrificed and I let the tents and Kin's bedding go, telling the soldier to stay behind and look after them, and follow us as soon as he could, with two more porters.   To delay any longer ourselves might have been fatal to the enterprise, and we had already wasted the morning, to say nothing of the previous afternoon.

As a matter of fact our soldier never came and we accomplished the journey without an interpreter and without tents.   He told Kin afterwards that he did not mind the beating, but was afraid of the snow and of mountain robbers; he had therefore taken the loads we had left behind straight down to Tsu-kou without troubling himself about our plight, and consequently he did not get the handsome bribe I had offered him.

It was two o'clock before we started for the pass, camping well up in the forest at nightfall ; Kin, who luckily had plenty of thick clothes with him, borrowing enough of my bedding to leave me rather cold without making himself much warmer.   And cold it certainly was now, though camp was pitched below 10,000 feet.   It was raw too, for the drizzle of rain which had set in towards dusk later turned to snow, and everything was so wet that it was impossible to keep up a good fire.

Plate XXXII

Francis Garnier Peak (14,000), at the head of the Chun-tsung-la

The men had selected three magnificent *Cunninghamia* trees for the encampment, and here we had three fires, one for Kin and myself, one for the three Tibetan porters, and one for the Lutzu porter who was now joined by a friend also returning to his home on the Salween, thus bringing the strength of our party up to seven. It is a curious fact that the Tibetans and Lutzu will not eat or sleep round the same fire, as I remarked on both journeys. Returning from the Salween we had as porters four Lutzu and one Tibetan, the latter always selecting the best place available, usually near me, leaving the Lutzu by themselves some yards distant; from which I conclude that the Tibetans look down on the Lutzu as an inferior tribe.

In spite of a thick mass of vines and other creepers hanging from the giant trees above us, it was impossible to place my bed under shelter, and a fine drizzle filtered through on top of me. I awoke at an early hour, feeling very cold. The water in the buckets was coated with ice and a hard frost lay on the ground, and no sooner did I get out of bed than my feet and hands were numb with cold. We were off before the sun topped the ridge, the dead leaves frozen stiff crackling merrily under foot, everything white with rime. Before long the frost-bound path was under snow which grew rapidly deeper, but though soft, it was not of sufficient depth to cause us much inconvenience till the last thousand feet up to the pass was reached.

The temperate rain forest now looked its best. For the most part the trees were evergreen conifers, firs with occasional clumps of *Cunninghamia* and other species, but down by the stream were birches, alders, and maples, marked as spots of gold, orange, and red which, in the light of the rising sun, seemed to fill the dark forest with a rich mellow glow pervading everything, and very beautiful. With the thaw, the leaves came whirling down in their hundreds, rustling softly, and, catching the first sunbeams, sent shafts of coloured light twinkling and dancing down the woodland glades. Immense skeletons of *Lilium giganteum* rose stiffly on either hand, some of them ten feet high and bearing a dozen or more big capsules which were slowly scattering their useless seeds, for so far as I could make out, this plant, though producing many thousands of seeds rarely sets a fertile one.   Here and there, too, clumps of terrestrial orchids caught the eye.

Towards mid-day the forest grew more open and the glare from the expanse of glittering snow became intolerable.   Far away to the east, through the mouth of the valley we had ascended, we caught a glimpse of the Mekong-Yang-tze dividing range, but there was not much snow on it.

At a broken-down shed we rested an hour for lunch and then entered the bamboo forest, at the head of which the steep climb to the pass commenced.   The snow had by this time lost its crisp surface, and as it grew deeper, the going became very heavy.   There were magnificent birch and fir trees scattered about here amongst the clumps of bamboo, as well as big rhododendron bushes.   A steep ascent through the forest brought us to a narrow gently-rising gulley where the soft snow had drifted to a depth of several feet, and scarcely even a shrub showed itself.   A biting wind whistled down from the broad saddle-shaped pass above, and the sun being now hidden behind the mountains the air was freezing.   Sticking up through the snow, in which we floundered knee deep, were the dead haulms of a varied alpine flora, rattling their gaunt frames gloomily in the cutting wind, and here I found another species of *Meconopsis*, of which I secured a few seeds.

The summit of the pass is dominated to the south-east by a magnificent rock pyramid, entirely snow-clad when we passed, rising about 500 feet directly above it; to the north-west it is continued as a low ridge gradually rising again as the Doker-la is approached.   Except in the immediate neighbourhood of the pass this ridge is clothed with forests of fir which, at their limit, show clearly enough in their mutilated forms the force and severity of the winds which sweep over the narrow col.

The Chun-tsung-la, as the Tibetans call it, is the easiest pass to the Salween in this region, being considerably lower than the Doker-la and considerably less steep than the Sie-la.   On Major Davies's map it is marked 12,900 feet, and judging by the vegetation and other comparative indications, I should put it down at about that, or slightly more. Nevertheless it is closed for two or three months in the

Plate XXXIII

The Chun-tsung-la, 13,000 feet

year, while the Rün-tsi-la on the Mekong-Yang-tze water-shed, at least 4000 feet higher, is open as long—a remark-able illustration of the difference in precipitation on these two watersheds.

All caravans going to Ba-hang cross the Chun-tsung-la, but though the path, as paths go out here, had been fairly good so far, the descent on the Salween side was bad, and I was glad I had not risked my pony's legs on it. The Tibetan pony is a sure-footed beast, but off the main road the more agile mule or donkey is to be preferred, at least as a pack-animal.

From the pass a fine view of the Salween-Irrawaddy divide was visible to the south-west, but there was very little snow on it; the ranges are evidently considerably lower to the south. To north and west the view was obscured by a high spur, over which peeped the twin snow-clad peaks of a single mountain, evidently Ke-ni-ch'un-pu already referred to.

There was less snow on the Salween side of the range, or it had melted more rapidly; but the descent was steeper and the mud treacherous. Cutting a path through dense bamboo brake growing on a steep slope never gives satis-factory results, for the great knobbly root-stocks are always left as formidable obstacles over which both men and animals are liable to trip, though high steps of rough hewn logs aid the wayfarer in some of the steepest places.

At nightfall we camped in the forest once more, and were soon settled round roaring fires, which we certainly needed, for we were higher up than on the preceding evening, and the cold was intense. The night was perfectly clear, and then suddenly the whole forest was lit up, the great birch-trunks glimmering faintly like silvered pillars, while through the tangle of branches the snow could be seen glittering on the mountains to the west. The full moon had risen.

Towards morning the cold made further sleep im-possible, and we were early astir. Descending into the deep valley below, we now found ourselves once more surrounded by forest almost semi-tropical in its luxuriance.

It will be remembered that, on the Mekong, Yang-tsa marks the northern limit of the rainy region, corresponding to a point a little north of Tsam-p'u-t'ong on the Salween. South of the last-named village there is no very high range between the Salween and the Irrawaddy, as I had looked clear over these snowless mountains from the low pass, and probably nothing effective between the latter river and the Bramaputra to prevent the rains sweeping right across from Assam.

Proceeding in a south-westerly direction from Yang-tsa, therefore, we were approaching the jungle region of the Salween, and crossing the high spur which had previously hidden that river from view, we now beheld it towards sunset from a point a little south of Tsam-p'u-t'ong.

Though the principal mountain chain receives a big rainfall, such is not the case with the low barrier ridges which flank it, blocked out by the parallelism of the tributary streams already referred to; and having crossed the deep valley which separated us from the last ridge, filled with magnificent forest trees amongst which some gigantic alders were conspicuous, besides a dense under-growth of *Impatiens*, ferns, and so on, we climbed up again and so out on to the bare bracken-covered hill side, where gigantic pine trees in turn formed quite a feature. And there almost at our feet as it seemed, though two or three thousand feet below us, flowed the Salween river.

To the west, backed by a long low bank of sullen purple clouds, opened the wide valley which forms an approach to India; a little higher up stream was visible the bare lime-stone cliff which marks the site of Tsam-p'u-t'ong, scarcely distinguishable in the gloom, and brooding over all rose the snow-clad peaks of Ke-ni-ch'un-pu, looking weirdly pallid in the evening light.

It was pitch dark before we reached Cho-ton, but the night air was now warm and balmy; the men soon gained admittance to a hut, and after a hasty supper I went to bed, delighted to sleep once more with only a few blankets as covering.

Though it had been cloudy towards sunset, the full moon rose into a clear sky, and when I looked outside in the middle of the night, I saw a beautiful sight, the brilliantly lit pale cliffs behind Tsam-p'u-t'ong standing boldly out against the heavy black masses of vegetation, and the river,

Plate XXXIV

Sunset over the Salween-Irrawaddy Divide: view looking south-west from the Chun-tsung-la

like frosted silver, winding between its high spurs. Towards morning the valley filled with dense mist to a depth of one or two hundred feet, and when we woke up we saw below us only a monstrous white river of cloud washing round the cliffs. This was not dispersed till the sun topped the high range we had just crossed, when it began to seethe and steam up, dewing the spiders' webs on the trees as it passed till they looked like gossamer. The same thing happened on the following morning, and I am inclined to attribute the phenomenon to the probable fact that the water of the Salween is at this time colder than the rocks of the valley, protected as they are from excessive radiation by vegetation, so that the air in contact with the river is gradually cooled down below the dew-point and precipitates its moisture in the form of a dense bank of cloud. I saw no such thing take place in the Mekong valley, where the rocks are largely devoid of vegetation and the radiation correspondingly more rapid; moreover the air is much drier there, since the high Mekong-Salween divide intervenes to desiccate it.

Cho-ton consists of about a score of huts built on a bluff a few hundred feet above the river; Cho-la, where we had stayed before, was visible on the flat river terrace at our feet, scarcely half a mile down the valley, and thither we proceeded after breakfast.

It was a glorious day, and I decided to rest here before starting on the difficult journey over the Sie-la; for after the cold of the mountains, what could be more delightful than to bask in the sunshine for a day! In the gullies several beautiful orchids were in flower, fields of buckwheat and millet covered the plain, and the women were sitting outside their huts weaving hemp cloth. Later I went down to the bed of the river, where the water was beautifully clear and green. Here the barometer read 24·96 ins.

Now Cho-la on the Salween, Tsu-kou on the Mekong and Pang-tsi-la on the Yang-tze are practically in the same latitude. In the river bed at Pang-tsi-la in September the barometer read 22·25 ins., and in the bed of the Mekong at Tsu-kou in November, 23·96 ins. These readings, of course, do not give the absolute altitudes of the river beds, but for comparative purposes they are I believe fairly accurate. Thus we see that the Yang-tze flows several hundreds of feet above the level of the Mekong, and the Mekong in turn several hundreds of feet above the level of the Salween. I had already suspected these differences of level, partly from the differences of vegetation in the valleys (though this was largely accounted for by rainfall) and partly from the readings of my barometer taken at various places in the three valleys, though not down in the river bed itself.

It is an interesting point, for it suggests the possibility of a great westward tilt of the country, as though the great mass of the Himalaya had caused a sagging of the earth's crust where these three rivers break through from the Tibetan plateau. Accurate barometer readings, however, at corresponding points in the beds of the Yang-tze, Mekong, Salween, and both branches of the Irrawaddy are much to be desired.

In Cho-la I met a third Chinaman who had crossed the mountains to the Irrawaddy as interpreter to Hsia-fu ; for having lived most of his life amongst the Lutzu he spoke their language and to some extent conformed to their mode of living. Hsia-fu had made him Tussu, or chief of the district for his services on that occasion, but the reward was a sinecure. He told me of takin on the Irrawaddy (wild oxen he called them), and described the Chutzu who inhabit those forests as a shy inoffensive people who tattoo their faces.

There were splendid apples here, now ripe, also juicy oranges and pomegranates. Tobacco, walnuts, gourds used for drinking-vessels, and magnificent pumpkins are also grown, besides the crops already mentioned. A common tree is a species of *Cornus* (probably *Cornus capitata*) with a curious aggregate fruit exactly like a strawberry, which is eaten by the natives ; this also grows in considerable numbers all the way down the Mekong valley from Tsu-kou southwards.

I tried to photograph some of the nicest-looking women in Cho-la, but met with scant success, for they ran into their huts, where I found them lolling round the fire smoking and tippling their warm soupy brew of fermented maize ; whereupon they would laugh slyly at me, but refuse to come

out.   On one such occasion a man handed to one of the
women a rosary, which she proceeded to count over, before
handing it back.   She had been almost persuaded to come,
but now she shook her head resolutely and returned to the
fireside.   The fates were against me, I suppose.   The men
were also generally to be found sitting round the fire smoking
and drinking, but no doubt there was little for them to do
in the fields at this season, so I forbear to call them lazy.

On November 8 we started back, following our old route,
and reaching Mu-la-t'ong in the evening.   Here I noticed
tomatos growing, but they were probably introduced by
the French priests.   The men did not want to go further
than Ba-hang next day, though we were there by twelve
o'clock, and I had lunch with the French priest.   We
therefore compromised.   They were to go on up the
mountain to a camping ground about half-way to the first
pass and leave me there, descend to Ba-hang for supplies,
stay the night, and return early on the following morning.
This arrangement would enable me to reach the first pass
in time to obtain an uninterrupted view of the Salween-
Irrawaddy divide before the clouds rose out of the valley, as
I imagined.

Camp was therefore pitched on the edge of the scrub
forest, composed chiefly of bamboo in the damper hollows,
of birch, rhododendron, and long grass on the dry slopes ;
the four Lutzu porters started back for Ba-hang, and the one
Tibetan, Kin, and myself were left in charge.

Meanwhile the weather had turned dull and threatening,
and the western ranges were buried in cloud, a chilly wind
began to blow, and we sadly missed the tents.   In the night
snow began to fall, but I resolutely hid my head under the
blankets and pretended to know nothing about it till it
melted and trickled inside the bed.   By the time the snow
had ceased the fire was quenched, but the men got up and
presently had it blazing again.

We had just finished breakfast in the grey of a bitterly
cold morning when the Lutzu arrived with supplies, and I
started at once up the steep grassy alps.   I need not have
troubled myself however.   The Salween valley as usual was
full of cloud, and the mountains completely obscured by
blinding snow-storms.   What was worse, the cold wind

blowing up the mountains was bringing it our way, and an hour later we were in the midst of whirling snowflakes.   On the other side of the pass a foot of snow covered everything, and I had great difficulty in finding my plants, which, when found, yielded few seeds, for they were mostly scattered and lying snugly beneath the snow blanket which was to protect them for seven months.   At this altitude (12,000–14,000 feet) on the Mekong-Yang-tze divide there was not a vestige of snow as yet, and gentians were still in bloom.

Half-way down the steep forested slope I was surprised to meet my friend Père Mombeig coming up, on his way to Ba-hang, a journey which he and his lightly-laden porters intended to accomplish in two days, for they had crossed the Sie-la that morning.   Thus they would cross from river to river in three days, which is remarkably good going. He brought reassuring news from Ssu-chuan, and advised me to waste no time in getting over the Sie-la, as the snow, which was now falling faster than ever, would make the pass difficult.

In the afternoon we reached the hut in the valley, but so different was its aspect that I scarcely recognised it as the same place we had slept at in June.   There was not a flower to be seen of all that glorious colour scheme upon which we had feasted our eyes only five months previously ; scarcely a vestige of that wealth of undergrowth we had waded through remained.   We made ourselves as cosy as possible round the fire in the draughty hut, myself tortured by the smoke, and all night long it snowed.   When we awoke next morning we gazed out on to a bitterly cold white world, on a scene essentially the same as that on which we had looked in June, and yet how changed !   All the wonderful alpine meadow which, gemmed with poppies, primulas, columbines, lilies, and many more, had occupied the valley bottom, had been utterly blasted by the autumn gales.   The alders and birch trees now flapped their long streamers of lichen dismally in the wind, icicles hung from every cliff, and the stream rolled its shrunken waters between ice-bound banks.

It was a long stiff climb up to the Sie-la and again rolling mists enveloped everything, so that we could obtain no view of the Mekong-Yang-tze divide to the east.   When

we crossed this pass in June, it may be remembered, there
was practically no snow on the precipitous Salween or south-
west side of the ridge, but now it extended from the valley
to the summit, the last few hundred feet—which is extraordinarily steep—being very difficult, for here the snow-drifts
were several feet deep. The Mekong side presented much
the same appearance that it did in June. Then, however, the
last thousand feet of snow was merely the unmelted remnant
of accumulated months, still deep and soft on the surface.
Now there could at most have been the accumulations of
only a few weeks' snow, but already it was several feet
deep, crisp on the surface, but not yet compact beneath.
The trail was obliterated, and consequently we floundered
knee deep at every step.

It had ceased snowing, but a howling gale raced over
the pass, whipping up the frozen surface of the snow and
filling the air with fine spicules of ice which stung the face ;
and what with these blinding clouds and the mist, the three
of us who were leading (for the Lutzus seemed uncommonly
inert) lost the trail completely and nearly slid over a
precipice. However, we ploughed across ledges of rock
and soon picked up the trail again, reaching the valley in
safety after a morning's hard work.

Judging by the amount of ice on the rocks, the Sie-la
had probably been under snow for some weeks, so that it
can be clear for little more than three months in the year.

That evening we got well down through the forest,
finding very little water in the torrent which had bothered
us a good deal in June ; indeed at one point it suddenly
disappeared beneath the shingle, flowing for several hundred
yards so far below the surface that not even a sound came
up to us, before it burst out again from its subterranean
channel.

We camped in a clearing at the foot of the last spur,
happily finding a small and much dilapidated shed which
Kin and I appropriated. It snowed heavily all night.
The Tibetan settled himself under the lee of the shed
where Kin had made the fire, and the Lutzus built their fire
fifty yards away ! Next morning the forest presented a
pretty sight, but it was very cold, and starting early, we
soon reached the top of the spur and saw the Mekong

below us through the rising mists. Before mid-day we were back at Tsu-kou where we found the rest of the caravan patiently awaiting us, Ah-poh being wild with delight at seeing me again.

Coming down the steep slope above the village I saw a party of Tibetans engaged in trapping vultures by means of decoy birds, which are led to carrion the moment the wild birds are seen, the men hiding meanwhile. When the wild birds, circling round the cliffs at an immense height, descry from afar the feast below, they come down to join the struggle for tit-bits, and the tame birds at once set on them; whereupon the men rush out from cover and despatch the intruders. The birds are thus caught for the sake of their feathers, which are sold in Tali-fu to make fans.

We spent two more days at Tsu-kou drying and packing the seeds we had secured on the journey, and on November 15 we finally started southwards for Wei-hsi in company with the deserter Gan-ton. The weather was warm and drizzly in the valley, but there was excellent duck shooting to beguile the time, and we made easy stages to Hsiao-wei-hsi.

The forests in the rainy region of the Mekong do not compare with the Salween jungles in variety of plants or diversity of form, lacking in particular just those features which give to the latter their essentially tropical aspect. Pines and oaks are conspicuous but, except in the deep gullies, the deciduous-leaved trees below very soon give way to fir forests above, so that the rain belt, which extends north and south for about a degree of latitude, is intermediate as regards its vegetation between the semi-tropical forests of the Salween and the xerophilous vegetation of the arid regions.

At K'ang-p'u I found a number of Lissus assembled in the Tussu's yamen, where I slept, and learnt that this chief rules over some 15,000 families, Chinese, Minchia, Moso, and Lissu from the mountains to the east. When I first saw them I instinctively thought of Mr Edgar's dwarfs, for the women, who made up the bulk of the party, were certainly dwarfish, with enormously developed legs. What was even more remarkable, several of them exhibited those

Plate XXXV

A Lissu woman of the
Mekong Valley

A Lama priest of Wei-hsi

unmistakable negritoid characters which had so struck me from time to time amongst the Tibetans and Lutzus. "Here," said I to myself, "are Edgar's dwarf slaves." But afterwards I was told that each family worked for the Tussu only five days in the year, and I was lucky to find them in the courtyard having their supper, a meal which consisted only of a greasy pottage such as is given to pigs in England.

These women were dressed in pleated skirts such as the Mosos wear, except that they reached only to their knees, the calves being roughly bound with puttees; but unlike the latter people they were ugly, dirty, and altogether grotesque, the only pleasing feature about them being the cloth head-dress covered with cowries. These are probably heirlooms handed down the family, for all my efforts to purchase one were unavailing though I offered its weight in silver!

Next day we reached Hsiao-wei-hsi, where I had supper with the French priest. This was as far as Gan-ton and his friends had arranged to come, though before setting out from Tsu-kou we had tried hard to persuade him to go with us either direct to T'eng-yueh by the Mekong valley, or to Wei-hsi; but the men were afraid to risk either themselves or their animals on the small road, except at a perfectly outrageous figure, and made ridiculous demands for the two days' journey on to Wei-hsi.

I therefore decided to be unreasonable too for once, and try to bring Mr Gan-ton to his senses; so for a start I refused to pay him till we left Hsiao-wei-hsi. He might suit himself, either go on to Wei-hsi at my price, or find me other transport.

We spent a whole day arguing over this; he came down in price and I raised my offer till we were within a shilling of each other, but I stuck to my principle and refused to concede the point. I was in no desperate hurry, but bored beyond measure; nevertheless I settled down in camp at Hsiao-wei-hsi with the appearance of being quite happy, collected plants, and amused myself. Then Gan-ton tried to get even with me by telling the few villagers we had secured as porters not to go with me, but when he found that I was adamant and that the French priest was powerless to persuade me—that, in short, he would get no money till I did go—he gave up that line. For in the meantime his animals had to be fed, so that my position, if unconstitutional, was entirely unassailable.

Next morning the porters who could not possibly be found the day before came in any number, and when the last had started, I paid off the smiling Gan-ton, feeling half inclined to pay off his companions and leave him out, as indeed he had himself generously suggested on the previous day, for as I reminded him, he still owed me two weeks' service. However I was content with my moral victory, the only one I ever gained over that astute linguist, and we parted the best of friends.

Two days later we reached Wei-hsi, where I learnt definitely that southern Yunnan was ablaze with revolution, the capital, Tali, and T'eng-yueh being all in the hands of the revolutionists.

No muleteers would listen to a proposal to take the caravan by the small road through the tribal country, so I decided that the safest course was to send it under escort by the main road, and myself go by the small road ; for my presence was likely not only to embarrass the few muleteers willing to travel at all in these troublous times, but also to invite attack from robbers or from marauding troops. Moreover it now became necessary for me to reach T'eng-yueh with all possible speed.

The next thing was to put this plan into operation, and I therefore called on my friend the T'ing-kuan, who promptly scouted the idea. But I was firm, and met his remarks on the state of the road, the length of the journey, and the number and ferocity of the brigands by saying that it did not matter, and anyhow I was going. Finally he gave in, promised to secure the services of two porters and provide an escort and interpreter, and bowed me out. Happily he was suffering from a frightful cough, and I took the opportunity of sending him round some chlorodyne with elaborate instructions, and so delighted was he that he in return sent me a parting gift of a fowl, a haunch of bacon, a tin of tea, and a packet of dried morells, which proved excellent eating.

CHAPTER XV

THROUGH THE LAND OF THE CROSS-BOW

WE left Wei-hsi-ting next day, November 23, our two caravans creating quite a stir in the little city as they clattered down the stone-paved slippery street and set out in opposite directions.

The main party, under the leadership of Kin, who with conscious dignity rode Beauty, comprised five pack-mules in charge of two muleteers, and an escort of three soldiers told off to accompany them across the Li-ti-p'ing, a somewhat desolate region of forest and bog, as already described, at the summit of the watershed, generally supposed to be infested with Lissu robbers. The soldiers would escort the caravan only as far as the Yang-tze, two days' journey hence, after which the care of my property devolved upon Kin, at least as far as Tali-fu. However, he was amply provided with silver, and carried my passport which, the official had informed me, would be respected by Imperialists and Revolutionists alike, did my caravan fall into their hands, so I apprehended no trouble on that score. My own party consisted of Sung; a yamen-runner who had charge of my local passport, with power to commandeer interpreters, escorts, and anything else he considered necessary for the safe prosecution of the journey; two porters carrying our bedding and a few cooking utensils; and finally Ah-poh, who was destined to create a great sensation in southern Yunnan.

As we went up the valley to the south-east we passed through several villages where groups of pretty Moso girls stood laughing and chatting in the sunshine; many people were hurrying into the market at Wei-hsi, some of them from a considerable distance, but all, in spite of their heavy loads, smiling and cheerful, with a word for the men and bright eyes for me.   Charcoal from the little forest settlements and earthenware pots were the chief commodities.

Right away up the valley to T'o-che, the last Moso village, where we stopped for lunch, rice-fields terraced the gentle slopes, so that here rice is cultivated well above 8000 feet, a somewhat unusual altitude.   It was good to see the unwieldy buffalos floundering through the liquid mud again.

At T'o-che we added a Pê-tzu interpreter to our party, it being his duty to find us quarters at the next village, which was inhabited by Pê-tzu and Lissu families, amongst whom he was well-known.   Had our official-looking party arrived unceremoniously without such a mediator to plead for us, the villagers might have received us with mixed feelings. As a matter of fact, however, we found that most of these people spoke a mangled form of Yunnanese, though even my men sometimes found a difficulty in understanding the more outrageous localisms.

After a brief halt at T'o-che we turned due south, still following up the Wei-hsi stream, the valley narrowing considerably.   Thickly forested sugar-loaf hills rose on either hand, but as evening drew in we emerged once more into open rolling country and here amongst widely separated patches of cultivation were scattered some twenty wooden huts.

The good people of Ching-k'ou-t'ou also belonged to the Pê-tzu and Lissu tribes, though they had for the most part adopted Chinese dress ; their huts, however, in one of which I was made welcome for the night, resembled those common amongst the tribesmen.   Maize was the staple crop, and the scattered nature of the village was obviously occasioned by the difficulties of cultivation amongst these undulating hills, rather than by any system of land tenure. Each family establishes its home on the virgin soil where there is a reasonable chance of successfully clearing the forest and cultivating the hill side.

After a sharp frost which covered grass and trees with glistening rime—for the altitude was over 9000 feet—we started off just as the sun appeared over the tops of the hills, and crossing the low watershed, descended through wooded hills to the little village of Ssü-shi-to, which, though comprising but twenty families, is widely scattered on the surrounding slopes.   Two interpreters had accompanied us so far, and two more were detailed to take us on to the next halting-place, but as a matter of fact these men were more in the nature of guides and escorts, due to me as travelling under official protection, than interpreters, who were hardly required at any stage of the journey.   Usually they were armed with big swords and cross-bows, but sometimes they carried only long spears in token of the official nature of their business.

I have already remarked that the Lutzu use the cross-bow, and the same is true of the Moso, Pê-tzu, Lama, Minchia, and Shan tribes with whom I came in contact ; it was, in fact, the universal weapon throughout the Mekong and Salween valleys as far south as latitude 25° 30'.   I do not remember to have entered a single hut between Wei-hsi-t'ing and Kai-t'ou, where there were not several of these weapons, each with its quiver, hung against the wall. Except amongst the Shans, who use simply a bamboo tube, the quiver is invariably an oblong box made from the skin of the black bear, containing two or more bamboo tubes, in one of which are kept plain arrows, in the other poisoned ones, the poison, or ' medicine ' as the Chinese with unconscious irony term it, being made from a species of Aconite.

The cross-bow has been regarded as the especial attribute of the Lissu tribe, or at least as eminently typical of them, but this is by no means the case.   Its simple structure, short range, and diabolical effectiveness mark it as emphatically a weapon made by a jungle tribe for jungle warfare ; hence there can be little doubt that it originated amongst the tribes of the river valleys between Assam and China, whence it spread eastwards.   The scarcity of birds in these valleys is no doubt partly due to the fact that every small boy carries a miniature cross-bow and shoots at everything he sees, much as the English country boy carries a catapult, and probably for the same reason—to amuse himself and become skilful in the art of killing things, rather than to provide food.   Boy is more or less of a barbarian everywhere.

From Ssü-shi-to two of my soldiers turned back, leaving

me in charge of their corporal, an excellent fellow who made himself both agreeable and useful ; my yamen-runner too did his work satisfactorily, so that the next three days passed pleasantly enough.   Descending some two thousand feet through pine forests, by a path down which not even a Yunnan mule could have scrambled, we reached Kûng-chou on a small river which here takes its name from this little village of ten or twelve Lama families, though no doubt it has several other names higher up, for geographical nomenclature is notoriously localised in China.   The cultivable area being confined to narrow strips in the valley bottom, the huts, instead of being widely scattered as in the mountain settlements, were bunched together above the shingle and boulders which marked the high water limits of the now shrunken river, so that the village looked even smaller than it really was ; indeed, it could have been comfortably tucked into a moderate-sized English farmyard. These people carry on practically no trade and cultivate almost everything they require, which is not much, though wheat, millet, maize, tobacco, hemp, ' red pepper,' beans, and turnips are all grown.   Still, I do not think the whole area of soil cultivated by the twelve families of Kûng-chou, boxed up in their narrow wooded valley, exceeded two acres.

Situated a few miles higher up the Kûng-chou river is the much larger village of Kow-shan-ching, where resides a petty mandarin with twenty soldiers under his command, probably keeping an eye on the salt wells, for on the following day we met a small salt caravan going down to the Mekong valley under escort.   Salt, being a government monopoly in China, is jealously guarded, and the most elaborate precautions are taken against smuggling.

On the way down the Kûng-chou river next day I shot a cormorant in mistake for a duck, to the intense delight of my village escort, who were presented with the spoil ; for though they assured me that it was excellent eating, I felt that they were more likely to appreciate the flavour of the ' old man of the water,' as they called it, than I was myself.   A few miles lower down, where we stopped for the night, the narrow valley, hitherto sparsely inhabited, widened out, and several large villages dotted the gentler slopes of the mountains.   The open forest of pine and oak,

which gives to so much of this country a withered and
forlorn aspect, was again replaced by terraced rice-fields,
and snow-clad mountains dimly seen through the clouds to
the west proclaimed the fact that we were not far from
the Mekong.

A large village called Pi-iu-ho, one of a group of three,
was our halting place, but though the population was entirely
Lama, there were—in addition to wooden huts with shingle
roofs—not a few houses built of mud bricks after the manner
of the Chinese, and roofed with tiles. Nor were the people
by any means ignorant of western curiosities, for while
taking a compass-bearing surrounded by a crowd of interested
Lama men, I overheard one of them instructing his more
ignorant companions to the effect that the instrument I held
was a watch. Possibly he did not know what a watch was,
but anything which is more or less round and has a needle
under a glass face, such as a compass or an aneroid barometer, is a watch in Western China, so that my friend's
erudition was quite astonishing for a Lama.

Major Davies considers it probable that both the Lama
and the Pê-tzu tribes are really Minchias under another
name, but this I did not find to be the case. They themselves emphatically denied that they were either Minchia or
Chinese—they seemed quite hurt at the mere suggestion—
and though I did not take down any vocabularies, I heard
both Lama and Minchia spoken, and certainly they were
not the same ; the Lama tongue, or what I heard of it, had
a superficial ring of Tibetan about it, though I do not for a
moment mean to suggest that there is any connection between
the two. But to point to any characteristics, either of
scientific value or of simple convenience which will serve
to distinguish the Lamas from the Minchias on the one
hand, and from the Chinese on the other, is a difficult
matter. As Major Davies remarks, the unconscious power
possessed by the Chinese for absorbing the ruder people
with whom they come in contact is quite astonishing ; only
the Tibetans seem able to resist—even to reverse the
process. But the peculiarity of the case lies in the fact that
there are scarcely any Chinese between Wei-hsi and
La-chi-mi, the inference being that the people themselves
are immigrants, and were thoroughly imbued with Chinese

ideals before they came into the country. This in fact proved to be the case, for several men, on being asked how it was they spoke and understood Yunnanese, replied that they came originally from southern Yunnan.

The Lama men have adopted Chinese dress, and gradually, no doubt, assumed Chinese manners and customs to a sufficient extent to destroy their own distinctiveness as a tribe. More remarkable still, they usually talk Chinese amongst themselves while vehemently pleading not guilty to the charge of being Sons of Han. Only on very few occasions did I hear men speaking Lama in ordinary inter-course. No doubt if the men dressed otherwise attention might be focused on differences of feature between them and the Chinese, and I suppose a trained anthropologist, who would know what to look for, would remark these without such external aids. There can be no question for instance that even the Panthays or Yunnan Mohammedans, who have intermarried with the Chinese for centuries, still retain a certain distinctiveness which makes them easily recognisable. But I must confess that these more subtle differences baffle me.

A few of the Lama men here, it is true, dressed only in a loose white cotton vest and drawers, but they seemed of a very poor class, and were not sufficiently numerous for purposes of comparison; they may indeed have been slaves.

With the women on the other hand I had less difficulty, though they too had adopted Chinese dress, or a very fair imitation of it. The fact that they did not bind their feet, while sufficient to show that they were not Chinese, did not of course prove that they were Lamas. However far the absorption of a tribe may proceed, the women seem rigorously to eschew this revolting practice, as witness the Manchu women. Unlike those tribes who retain their native costume, the Lama women wear little jewellery; small silver ear-rings and bracelets being quite inconspicu-ous after the heirlooms one is accustomed to see pendent from the head of a Moso woman. The hair is worn in a queue, at least amongst the unmarried girls, just as it is with the Manchu women, whereas amongst the Chinese that emblem is—or was—confined to the male sex.

In spite of these confusions I usually found that by carefully looking through the women folk of a tribe, however much they might superficially resemble the Chinese, it became possible after a time to fix on a distinct type which occurred over and over again, a common factor for all the tribal features, so to speak, and this I found to be the case both with the Lama and Minchia women. Moreover a peculiar little cap, rather after the style of a military forage cap without the dent in the middle and ornamented with large metal studs round the edge, was fashionable amongst the Lama and Minchia girls. Set jauntily on the head with a metal-studded collar to match, it looked very nice, and some of the Lama girls were by no means ill-favoured.

On the following day we reached T'o-yie on the Mekong, exchanging the somewhat cloudy weather which had dogged our footsteps while traversing the mountains for brilliant sunshine once more. Snow was visible on the Mekong-Salween watershed to the west, but such peaks as could be seen were not conspicuously high. As hitherto, each village of any size through which we passed furnished two or three men to act as escort as far as the next village, and in this way we picked up some curious specimens of humanity, armed with the most ferocious-looking swords and cross-bows, these temporary followers already having numbered no less than fifteen.

The fine hot weather of late autumn had now lasted for some time and the valley in consequence presented a decidedly desiccated appearance. The mountains of purple shale or grey metamorphic rocks sparsely covered with tufts of spear-grass and scattered pines or oaks were cleft by narrow-lipped gullies of great depth, suggesting furious rains in the mountains above and a very slight rainfall in the valley itself. Down towards the river the gentler slopes were extensively cultivated, though villages were few, and the high banks were fringed with shrubs, mostly evergreen, which hid the river from view. Here and there grew dense bushes of prickly pear. We were again in an extraordinarily dry region, the appearance of aridity increasing rather than diminishing as we continued southwards, and though hardly comparable to the true arid region

of the Mekong north of Yang-tsa, it was plain that there
was not here the summer rainfall which we got at Tsu-kou.
The extreme narrowness of the Mekong rift, and the greater
elevation of the mountains further east acting as rain-
conductors, probably accounts for the monsoon here passing
harmlessly over the Mekong and drenching the plateau
to the east. As above Yang-tsa, the deep ravines cut out
by the mountain torrents demanded the most exasperating
détours, and the river, frequently interrupted by rapids, was
rarely navigable even for ferry boats. Rope bridges occurred here and there, but were always of the annoying
kind that landed one automatically over the middle of the
river, after which gymnastics were required whether one
decided to go on or turn back, the only alternative being
to stay there.

In the evening we reached the insignificant village of
Ching-p'an, where I slept in an extraordinarily smoky little
hovel. Everybody bustled about to garnish up the place
as much as possible ; three planks across two forms con-
stituted in turn bedstead, supper-table, chair, and wash-
hand-stand, for my apartment was so small we could get
nothing else inside, even if the resources of the establishment had run to it, which they did not.

On the following day we were due at Shi-teng, whither
my passport from Wei-hsi entitled me to travel under official
escort, which really means at everybody else's expense if
one is so minded. At Shi-teng I should have to obtain a
new passport from the local official, for no Chinese official
can give a passport for territory under the jurisdiction of
his superiors, though he can give a through-passport for
territory administered by a mandarin of inferior rank. We
had now reached the limit of the Ting of Wei-hsi's command
in this direction, both Shi-teng and La-chi-mi being under
the Lichiang-fu official.

However, when we stopped that morning at a small
village for the mid-day meal, I learnt that the Shi-teng
official had just arrived on his way up the valley, and our
interview took place immediately. As always when one
proposes to travel in unadministered territory, the official
found excellent reasons for requesting me to rejoin the
main road at Chen-chuan, which I could have done by

making a journey of four or five days duration over the mountains to the east; but having politely absolved him from the necessity of guaranteeing my safety, I adhered to my original plan, observing that if he could not give me a passport I would go without. The official, full of alarms, still hesitated, but on my rising to intimate that the interview was at an end, he now begged me to be seated again, and ordered his writer to make out a passport for La-chi-mi, at the limit of administered territory southwards. Becoming quite affable once the Rubicon was crossed, he took me out to admire his pony, shelved the matter which was worrying him, and spoke of the weather.

So much time had been wasted already, that by the time we had refreshed ourselves and started once more the sun was sinking behind the mountains. It was indeed nine o'clock before we reached Shi-teng, but the moon, seven days old, gave us sufficient light to see the road save in the deep ravines, and we found comfortable quarters at an inn.

I had hoped that the business of interviewing officials was now at an end, though I must confess to have always enjoyed these social engagements, which I regarded in the light of gratuitous Chinese lessons. It was always a relief to hear an educated Chinaman speak after wrestling with the streams of patois poured forth from the mouth of the ordinary coolie or clodhopper, and though of course no mandarin ever dreamed of speaking slowly simply because I prefaced the conversation by saying I did not understand Chinese very well, yet I generally managed to understand the drift of his remarks, and in any case always got what I went for, which was the chief thing.

The Chinese rustic is an absurd person, with the prevalent idea of his class in all lands that if you do not understand what he is saying, he only has to say it a little louder and you cannot fail to comprehend; consequently if you intimate that the full purport of his remarks has been lost on you, he immediately begins to bellow like a bull of Bashan, confident that you are only a little deaf, not ignorant of his language; and since in his eagerness he pours out the verbiage faster than ever, inoffensive people trying to get a cheap lesson in the language are apt to suffer.

However, I was now anxious to get on with my journey, and well knew that each meeting with an official meant at least half a day lost; consequently I was not filled with joy to learn that another mandarin had arrived in Shi-teng on the previous day from the capital. This gentleman visited me first thing in the morning, but finding me at breakfast, paid his compliments and retired; an hour later I returned the call and was most kindly received. The only annoyance was the delay of changing men, my two faithful followers from Wei-hsi now finally leaving me, so that others had to be procured at the orders of the official. It was mid-day before we got away.

Shi-teng is a village of some importance, boasting ninety families and a garrison of twenty soldiers to attend on the mandarin; it was the last place of any size where I came in contact with the Lamas. Shortly afterwards we left the Mekong and striking eastwards up a small valley were once more closely invested on all sides by mountains. Near the hut where we took up our quarters for the night I had the good fortune to shoot a snipe; the commissariat had not been working very smoothly since leaving Wei-hsi, for it was difficult to procure anything except eggs, the most trying part being that I could get nothing with which to make bread save buckwheat flour; and buckwheat cakes are far worse than dog-biscuits. On November 29 we started early, in the hope of reaching La-chi-mi by nightfall, for we had already lost much valuable time and it was by no means certain that I should reach T'eng-yueh ahead of the main caravan. Continuing up the valley we held on a southerly course parallel to the Mekong, which was now hidden behind a high range. There was a good deal of cultivation near here, and huts, though widely scattered and usually isolated, were frequent. In these sequestered nooks the Lamas seemed to retain their own habits and customs to a greater extent than did their fellow-tribesmen in the Mekong valley, and spoke their own tongue almost to the exclusion of anything else, though a few of the men could manage a little Chinese. The women, I had reason to believe, could not, for having got behind the guide and subsequently taken the wrong path, of which there were any number to choose from,

I found it impossible to make myself understood by several buxom-looking lassies I met, who were highly amused in consequence.

Crossing the low spur at the head of the thickly wooded valley we continued southwards, halting for lunch at the first collection of huts we encountered, four or five of them close together. Their doors stood invitingly open, the fires burnt brightly, but of inhabitants there was no sign—not even a dog barked. Either the place was deserted for the day, or we had stumbled across a village of the dead. However, the men, nothing daunted, at once began to forage round for supplies, though five minutes search resurrected nothing more nourishing than a bunch of turnips, which we fell valiantly upon and quickly destroyed, being very hungry. Later they found a solitary egg which was handed over to me, and meanwhile we had put the rice on to boil and had made some tea, while I turned my attention to the more immediate resources of the supply-box, in which however I discovered nothing but a few remaining slices of bacon. There was not even any bread for our meal, which was still further complicated by the fact that I had to cook it myself, Sung having foolishly got behind the guide and lost the way, as I had myself done earlier in the morning. He did not indeed reappear till the middle of the afternoon, when we suddenly perceived him toiling up the precipitous slope of the deep valley immediately below the path, halting every few yards to hold on to a pine tree. He was eating pea-nuts as though nothing had happened—it was five hours since we first missed him—and sent me into fits of laughter by relating with an injured air in his slow quiet way how he had been told to follow the road at the bottom of the valley.

Meanwhile the men had made themselves quite at home in the hut, turning everything upside down several times over in their eagerness to find buried treasure, or something interesting to eat. I left some money in payment for what we had taken however, being very careful to go out last and shut the door, as I knew well enough that if either of the soldiers saw what I had done they would not rest till they had got possession of that money. In fact the idea of paying for a thing when there was no occasion to do so would have struck them as so entirely novel a proceeding that they would have spent quite a long time explaining elaborately that as we required food and happened to have found it in a deserted hut, and as no one was there to accept payment in return, the only just and reasonable course was to take what we could find and be thankful ; and furthermore that the villagers would be only too pleased and proud to give the English excellency whatever he required—and incidentally of course whatever his men required also. The uneducated Chinaman can never see further than the end of his nose ; sufficient for the day is the profit thereof.

We had not proceeded fifty yards from the hut when I remarked that one of the soldiers was carrying in his hand a copper kettle, which, though I had never previously noticed it as part of his personal luggage, looked strangely familiar. Feeling suspicious, I asked him where it came from, and somewhat crestfallen, he replied that he had brought it away from the hut. I was so annoyed at this brazen confession that I hit him in the face with my fist, whereupon clumsily tripping over his rifle he fell in a heap to the ground, and lay like a half-empty sack of corn, bleeding from a cut lip. I now ordered him to take it back to the hut immediately, and convinced that he would only help himself to something else by way of compensation if he could, I escorted him myself, prodding him in the back with my gun and threatening to shoot him if he ever did such a thing again. Meanwhile he was plaintively apologising for his conduct and making ridiculous excuses, the chief of which was that he did not think I would mind !

It was indeed owing to this very fact that the soldiers always endeavoured to take advantage of the immunity which they fondly imagined my presence gave them to get something for nothing, that I refused escorts on every possible occasion when travelling amongst the Tibetans and tribesmen. The Tibetans, I believe, hate and fear the Chinese soldier, and consequently look with suspicion upon anyone travelling under his protection ; the Chinaman in turn fears the Tibetan, man to man, but he also despises him, and lets slip no opportunity of showing it.

A tribesman would hate me far less for looting his

house of half its contents than he would for allowing a Chinese soldier of my party to take a single copper kettle —at least that is my opinion. The incident closed for the time being with the return of the kettle to its native fireside, but what at first sight seemed to be the sequel was rather curious.

It was impossible to reach La-chi-mi that day, for the new moon had already set and it had become very dark in the forest. The stars shone brightly in the clear sky, and a cold wind blew up the valley, so that we gladly stopped at the next village, situated in a clearing on the edge of the pine forest. The Minchias—the first specimens of the tribe with whom we had come in contact—received me kindly, and made all necessary arrangements for my comfort, though the dogs, being extremely jealous of poor Ah-poh, were not at all friendly. The huts as usual were of wood, with slat roofs kept down by stones, and the women, except for a tall dark blue turban standing up on the head like a pudding-cloth, much after the manner of the Shan turban, were dressed in the orthodox fashion. There was nothing Chinese in their appearance however, for they were one and all fine strapping wenches, though not conspicuous for good looks, their large vacant faces putting one in mind of suet dumplings. Cross-bows with quivers of bear-skin, hung in every room.

On the following morning we had only a ten mile walk before us to La-chi-mi, and before mid-day we stood at the summit of the hill looking down into the deep and narrow valley, where several hundred clustering grey roofs announced the presence of a big village. Away to the west, through the broad mouth of the valley, we caught sight of the Mekong-Salween divide looking very blue in the haze, its lower slopes darkly chequered with forests, its wavy crest lightly painted with snow, silhouetting it strongly against the deep turquoise sky.

For some minutes I stood looking at the scene, and then with something of a start noticed for the first time that my two soldiers, having quietly dropped behind, had disappeared altogether, without a word of explanation, a circumstance which the yamen-runner presently explained by saying that they did not feel themselves worthy to enter

La-chi-mi! Probably they were not in any case, but I mentally attributed their behaviour on this particular occasion to that more metaphorical form of unworthiness best understood by the Chinese themselves, and commonly translated 'face.' It did not in the least matter that they had deserted—I was glad to be rid of them so cheaply— but I had no sympathy with the principle involved. It occurred to me that, fearful of being reported at the yamen and beaten for their conduct on the previous day, they had considered it advisable to accompany me no further, even though, by quietly slipping away, they lost all chance of receiving the small emolument it is customary to give soldier escorts, according to their merits. But I was mistaken, and I might indeed have been able to put quite a different interpretation on their strange action had I listened to the whispered conversation they held with a group of villagers some miles back.

At last we descended the thousand feet or so into the deep hollow, the glimpse of closely-packed roofs rapidly giving place to a more extensive view of the village, where Ah-poh and I created a good deal of astonishment amongst the Minchia and Chinese population. Having deposited my things in an inn and performed a simple toilet according to my resources, I repaired at once to the yamen.

CHAPTER XVI

THE REVOLUTIONIST OCCUPATION OF LA-CHI-MI

A BUSY little village, La-chi-mi owes its importance directly to its salt well and indirectly to its position at the junction of three trade routes, one going south down the Mekong to the main road at Yung-chang-fu ; a second (and a very excellent paved road too, at least near La-chi-mi) going eastwards to the main road at Chen-chuan ; and a third northwards up the Mekong to the main road at Wei-hsi. We had ourselves followed the last mentioned to some extent.

The village is situated at the head of a deep ravine, some miles from the Mekong, and is closely invested on all sides by rather bare mountains. It boasts several small temples, one in an excellent state of repair, while an arched and roofed wooden bridge, which I found in course of construction, over the stream where it enters the gorge will when finished, be decidedly picturesque. The population of rather over two hundred families is chiefly Minchia, but the numerous shops and tiled roofs give a distinctly Chinese aspect to the village, enhanced by the manners, customs, and speech of the inhabitants. On the other hand, many of the houses are built of mud bricks and roofed with wooden slats, and the women present those few but obvious differences from the Chinese that we have already mentioned. The yamen was a quaint old-world little place with two ancient cannon guarding the entrance at the head of a flight of steps, and numerous evergreen shrubs and dwarf trees standing in wooden tubs round the courtyard. Two dove-cots were hung against the wall and the soft cooing of doves sounded pleasant in the bright sunshine which flooded the yamen; on the roof a tame monkey was disporting itself at the end of a long chain.

I noticed with a vague sense of surprise, without attaching any particular importance to the fact, that the courtyard and surrounding rooms were full of armed soldiers, thirty or forty of them, big men dressed, not in the ragged official jacket characteristic of the Yunnan 'brave,' but in stout blue cloth with turban to match; and every man carried a bandolier and magazine rifle.

The official received me kindly, taking both my hands in his as he had perhaps seen the Europeans in Yunnan-fu doing, but I remarked that he seemed greatly agitated. However, he promised to find the requisite men to accompany me on the morrow, and I returned to the inn, puzzled.

Here I learnt that the revolutionists had that very morning occupied La-chi-mi, and furthermore that the main body, comprising about 250 men, were to enter the village next day. This, then, was the meaning of that strange flag over the yamen which I had seen the men gazing at from the hill top, and of the armed soldiers in the courtyard, the advance guard of a flying column.

In the afternoon I decided to pay a visit to the salt mine, which was not at this season in operation for reasons which will presently appear; the only available salt now being that which had crystallised on the numerous gutters and their supports used for the purpose of conducting the brine to the huts where evaporation is carried on; and a number of men were busily engaged in chiselling off these incrustations. While following up the gutters with a view to finding the position of the mine a friendly operator hailed me, and fetching a lamp from an adjoining hut, offered to act as guide, a proposition to which I gladly assented.

Proceeding up a narrow gulley just clear of the village we came upon the entrance to the cave high up on one bank, and lighting his oil lamp, my guide opened the doors and led the way into the cliff. The tunnel was about five feet high, in width rather less, and between the baulks of timber, like railway sleepers, with which the sides were shored up, an efflorescence of salt glistened on the moist red clay.

These side timbers, which occurred every yard, sloped

inwards several degrees and were kept in position by another set of timbers, one wedged between each pair, the latter serving also to support the roof, which was consequently even more contracted in width than the floor. Along the side of the tunnel ran a gutter down which flowed the brine when the pumps were at work, and being now dry, the sandy bottom, even in the feeble light of the oil lamp, glittered as though covered with hoar-frost.

After walking in a cramped position a distance of a hundred and twenty paces along a fairly level floor, and acquiring a severe crick in the neck as a result, we reached the first pump. From this point two passages branched off and we now descended rapidly by a steep flight of roughly-hewn steps into the bowels of the mountain. Meanwhile the height of the tunnel had contracted to four feet, and the atmosphere was close and hot. At the foot of the steps we reached a second level and espied close at hand a big pool of brine into which dipped the other end of the wooden tube we had just seen above. This tube, the bore of which was about four inches in diameter, was in reality the barrel of the pump, and the reason why the pump was at present out of action now became apparent, for the depth of brine was not sufficient to cover the lower end of the tube.

From this level, where there was a second pump, another steep descent took us to a still lower level where there was another pool of brine and yet a third pump communicating directly with the main brine reservoir, sunk deep in the heart of the mountain as it seemed; but to this, the real brine well, which received percolating contributions no doubt from vast distances, the two pools above being artificial reservoirs dependent on the pumps, we did not descend.

Thus the brine is pumped from level to level by relays of pumps, each operated by one man, until it reaches the gutter in the main tunnel and is run off to the several evaporating houses.

We now returned to the upper world as I wished to ascertain how the pump worked—a very simple business, seeing that it consisted only of a barrel and plunger. This barrel was made, not of bamboo as one might have expected, but of a hollowed tree-trunk, perhaps pine, though the bark had been stripped from the outside and it was bound at frequent intervals, no doubt to prevent the wood warping and cracking. The plunger consisted of a thin bamboo about six feet in length, near the end of which was fastened a circular leather apron, slightly larger in diameter than the tube. To the circumference of the apron were attached light guy-ropes, these being drawn down and secured at one point to the handle, just like the ribs of a half-open umbrella; and save for the fact that these stays were not rigid, that is exactly what the plunger most resembled.

Evaporation of the brine is carried on by a number of families living in miserable huts clustered near the mine at the lower end of the village. Four or five iron pans, each capable of holding a block of salt weighing not less than seventy-five pounds, are let into a mud range, the fires are lighted underneath and the process goes merrily ahead; but at this time no evaporation was in progress and consequently everything—walls, floor, range, pans, and conduits—was encrusted with salt. In La-chi-mi salt is sold at the rate of three taels of silver per hundred *kin*, equivalent to a shilling for eighteen pounds—certainly not a high price. The mine supplies Chen-chuan, Tali-fu, Yung-chang-fu, T'eng-yueh, and even distant Bhamo.

While strolling outside the village in the evening I met a patrol, who with lanterns swinging merrily, were almost running along the paved road which goes eastwards over the mountains.

On the following morning no men were forthcoming and I occupied myself with watching developments. There was little excitement in the sleepy village, for though many people were in the narrow street, all the shops were open. Presently the cry arose, "They are come!" and all who were not already in the street surged to the shop-fronts—it was the only indication of excitement while I was there. Coming down the paved street from the direction of Tali-fu, seven days' journey to the south-east, was a well-dressed man on horseback, preceded by some half-dozen foot-soldiers escorting as many pack-mules. This was Captain Li. Evidently the party, who had come straight up the Mekong valley from Yung-chang, were in constant communication with a second party who had proceeded northwards by the main road, their object presumably being to converge simultaneously on Wei-hsi, and thence marching northwards to Batang, to join hands with the Ssu-chuan revolutionists. From time to time small patrols of three or four men hurried out of the village, talking and laughing, most of them going eastwards in the direction of Tali.

As there seemed no immediate prospect of getting away unless I stirred things up in person, I repaired once more to the yamen and was again cordially received by the much-harassed official, who looked even more apprehensive than on the previous day. The new arrivals, seated at breakfast in the courtyard, their pack animals standing by, eyed me curiously, but said nothing. As before, armed soldiers stood about in groups chatting, strolled casually into the room to join in the conversation, and made themselves very much at home ; the yamen was at their mercy, the official virtually a prisoner.

The unfortunate man said that all his soldiers had deserted and hence he could not possibly make arrangements for me, or find men to go till the morrow, but that if I would only be patient and wait one more day, I should certainly be able to start next morning ; so hard did he beg me to stay that I began to think he felt a certain amount of security in the presence of an Englishman, who might at least be relied on to see fair play. But as a matter of fact the official had little to fear, and in any case I could have done nothing to help him, being helpless myself.

Procrastination is the thief of time however, and I insisted on starting at once with two porters, for, as I tried to explain, should my mules reach T'eng-yueh in advance, it was quite possible that since they had farther to go the Europeans there might fear some accident had befallen me.

Returning to the inn I found many of the soldiers standing about in the street, chaffing with the inhabitants, buying oranges from the children, and drinking tea at the tea-shops. They seemed to be plentifully supplied with money and paid for everything they bought, which was all these village people cared about. They became interested in my shot-gun, and requested me to show them how it worked, whereupon we fell to chatting.

At last my porters appeared, and shortly before mid-day

we got off, now left to our own resources since we were no longer under official protection.   We had scarcely entered the ravine below the village, when there came a shout from behind and Captain Li, surrounded by a little knot of retainers running fast in order to keep up, rode rapidly by. Seeing me however, he reined in his pony, and very politely enquired my business and whither I was bound.   I answered his questions, wished him God speed, and he dashed ahead to meet his troops who were on their way up from the Mekong.   Half an hour later we fell in with the main body of the revolutionists—but what a change!

They formed none the less a striking picture winding in single file up the narrow ravine, rather more than two hundred of them, though the majority were baggage coolies. There were Lissus and Minchias, sturdy little tribesmen with muscular chests and swarthy complexions, often ferocious of aspect, carrying *dahs* thrust jauntily through their belts and a huge three-foot cross-bow slung over the shoulder ; Yunnanese 'braves' in tattered official tunics, carrying rifles, most of them muzzle-loaders ; men with extraordinary looking guns of abbreviated length and immense calibre, after the pattern of a blunderbuss ; others carrying scarlet banners and long trumpets ; and half-a-dozen men, including my friend Captain Li, riding ponies.

How strange they looked, these poorly-clad tribesmen armed with *dah*, cross-bow, and arrow-case of black bear-skin, each carrying a load of no mean weight on his back by means of a strap passing over his forehead.   Occasionally the trumpeters put the long trumpets to their lips and throwing back their heads, made the welkin ring ; and on they hurried, banners fluttering, till they wound out of sight in the narrow gorge.

What did it all mean ?   What could it mean but that Li was inciting the tribesmen to rise and fight for—what? An independent Western China ?

Not long afterwards, when I was behind my men, a horseman accompanied by the usual armed sycophants on foot rode out from a village and hailed me gruffly. He appeared to be a person of no consequence what-ever, and resenting his manner, I took scant notice of him, whereupon he spurred forward and barring the way, demanded roughly enough my business. I of course gave him the required information though with no very good grace, adding that Captain Li had already spoken with me, which seemed to satisfy my interlocutor. "Oh! you have come from La-chi-mi! That's all right then!" he said, blustering, and straightway returned to the village. He was a person of no account and apparently wished to convey quite the opposite impression by offering me the only offence I received throughout the journey.

It was evening before we reached Ying-p'an-kai, a village of a hundred families fairly divided between Minchia and Chinese. Here I found comfortable quarters in a temple on the hill side, and amidst the pungent odour of burning incense made up my bed in one corner under the gaze of a galaxy of ferocious-looking gods.

One can usually prevail on the authorities of a Chinese temple which, as in this case, is sometimes tended by a layman, to give one shelter for the night, or even for the week-end; indeed many spots to which Europeans living on the coast or in the interior annually resort have no other accommodation than that afforded by a Buddhist temple, which is gladly put at their disposal by the bonzes for the sake of a little gain. But certainly they might just as well be put to such service, since they are no longer houses of prayer, Buddhism having no hold on the people of China; and this fact the Chinese have themselves recognised for some time by transforming many of the disused temples into schools and barracks. As a house of refuge from the curiosity of the Chinese a temple is infinitely superior to an inn, besides being as a rule both cleaner and more commodious. In this instance, however, the janitor demurred and remarked that since I had no passport, he was not going to let me sleep in the temple. Sung floundered on gaily, but he was too meek to make much impression on a really truculent man such as we had to deal with, and feeling rather incensed at the didactic tone he chose to adopt towards us, I now chipped in, saying that it was none of his business whether I had a passport or not and that I was certainly going to stay the night.

The man seemed rather surprised to learn that I had been following the conversation silently, but he now became entirely obsequious, carried my things in, made up the fire, prepared tea, and played the host as only a Chinaman can. In accordance with immemorial custom he covered his retreat by shouting to half-a-dozen people to do things immediately, and remained extremely affable till I left.

I was scarcely dressed next morning when there arrived in hot haste a runner bearing cards from Li, on which it was set forth that Mr *Wha*, who was on his way to the Consulate at T'eng-yueh, collecting plants during his travels, was to be accorded every facility for the successful prosecution of the journey, was on no account to be interfered with by the villagers, and was further to be furnished with escorts from village to village, to see that these orders were carried out. And in proof of the respect in which Li was held, whether because the tribesmen sympathised with the revolutionists or more probably because they represented the most concrete form of authority with which they had ever come in contact, I may say that wherever I presented Li's cards I was accorded the best the villagers had to offer.

There is, in addition to the ferry, a single-rope bridge across the Mekong at Ying-p'an-kai, whence a foot-path leads over the mountains to that region of the Salween where dwell the 'black' Lissus; but I was not recommended to try it.

We now continued our journey down the dreary valley, alternately toiling over high spurs round which the river, far below us, swept with a dull roar in big S-shaped curves, and descending to the bottom of deep ravines cut out by the torrents. Vegetation became less and less abundant, a horrible dryness pervaded everything, penetrated everywhere, and poor Ah-poh, with head and tail hung low, trotted along behind us in a very dejected mood. Perhaps he was thinking of the snows and the cruel winds of his native plateau, his spirit obsessed by the dun-coloured world and blue roof which day after day he saw before him.

Whole mountain sides of red shale showed not a vestige of life, and where vegetation clothed the nakedness of the rocks, it consisted almost entirely of a terrible grass which thrust its awned seeds into anything that came against it.

Once they obtained a purchase, which was not difficult seeing that they were provided with needle points, these abominable contrivances, being prevented by barbs from retreating, rapidly worked through one's clothes, and having reached the skin, the point would penetrate to a depth of a millimetre or more, causing the victim infinite misery. Yet so small were these little instruments of torture—a wicked and successful provision of nature for the distribution of the plant—that they were extremely difficult to extract even when their exact position had been discovered, which was none too easy a matter. In the immediate neighbourhood of the few poor villages we passed through, the vegetation was a little more abundant and varied. Hedges, for instance, of a tall cactus-like *Euphorbia*, its inconspicuous red flowers just blossoming, and one of the numerous trees (*Aleurites*) which yield vegetable oil, were particularly noticeable in this land of rock and grass.

We reached the village of Hsiao-ta-chu in the middle of the afternoon, and the porters, wishing to halt for the night, spent two hours over their meal in order that they might urge darkness as a plea for their laziness. However, I had already intimated my intention of going on to the next village, and the only result of their obstinacy was that we started at sunset to do another six or eight miles. The road was not easy and it was some time before the moon, appearing over the summit of the ridge, lighted up the valley sufficiently to point the way. I was myself thoroughly exhausted by this time, scarcely able to drag one foot after the other up the mountain path, and it was an immense relief when at last we heard the barking of dogs. Struggling up the hill side we came suddenly upon a collection of Lissu huts, and the men soon found some good people who, in spite of our late arrival, were willing to put up the strangers for the night.

Continuing down the left bank, we reached the ferry at Lo-ma-di next morning; the region was as arid as anything we had seen on the previous day, but desolate as it was, the magnificent weather made it more bearable than it might otherwise have been. Wherever cultivation is possible rice-fields terrace the sides of the valley, and we also came across tobacco, the castor-oil plant, cotton, and hemp in this region, though in meagre quantities. A very quiet stretch of water contracted to thirty yards in width is utilised for the ferry at Lo-ma-ti, where we crossed to the right bank; advantage is also taken of the narrows to sling across a two-way rope bridge, the third we had seen since leaving Ying-p'an-kai. Judging by the paths up the ridge on the other side of the river to which these bridges lead, and the several parties of Lissus we frequently saw ascending, there were no doubt many Lissu villages hidden away in the mountains at no great altitude.

At Lo-ma-ti—which name seems to be attached chiefly to the ferry, for I saw no village to speak of—the olive green water flows between steep banks composed of vertically-tilted slate rocks, with a sandy shore below, the latter being of course completely covered at summer level; but quiet as are these narrow reaches of deep water, they are frequently interrupted by boiling torrents smothered in white foam.

The ferry-boat consisted of two dug-out canoes lashed together and the passage was effected with very little trouble, two men paddling us across in a few minutes. Furious gusts of wind were raging up the valley at the moment, blowing clouds of fine sand in our faces, but this ceased as suddenly as it had begun and all was peaceful once more.

Towards evening we reached the Minchia village of T'u-wau, and no sooner was I established in the best room which the Tussu's house afforded, than there came an interruption. Hearing loud exclamations in the courtyard I went outside and saw ranged up in front of the house five fighting men, each armed with *dah* and immense cross-bow. They were informing the keeper of the Tussu's yamen (the Tussu himself being absent) that the Lao-wau Tussu was about to arrive, and rooms must be immediately prepared for him and his followers. These men, Shans by their appearance, were the advance guard of the chief's party. Now all was bustle again, and half an hour later the Tussu himself arrived, with a body-guard of about forty men, Shans, Lissus, and Minchias, all armed to the teeth. Hearing that I was at the yamen he sent in his card, and a little later called on me in person, accompanied by two or three of his friends. The chief, who was in Chinese dress and spoke that language fluently, was extremely affable; he told me that he was on his way to La-chi-mi to join Li, and having given me instructions as to the best road to follow, took his leave.

Next morning I found that he had given the necessary orders for my receiving safe conduct to Lu-k'ou on the Salween, where another chief would take over the responsibility for my protection, though as a matter of fact I had no more escorts after leaving T'u-wau, and no trouble of any kind with the people. In my experience it was the rarest exception for the tribesmen of the border country, or the mountain people of Western China, to be anything but friendly towards me.

The Lao-wau chief had given instructions that relays of villagers were to carry my two loads from village to village, each of which was to furnish in addition one or two men as escort. Consequently a good deal of time was wasted on the road, as each village we arrived at implied a delay of quite half an hour while the village headman hunted up the requisite number of men—or women, as the loads were usually borne by the fair sex. As for the escorts, I never waited for them to put in an appearance at all. Previously I—or rather Sung—had engaged porters to accompany us for several days, so that delays of this nature had been less frequent; but under the new conditions of travel it was only when the villages lay far apart that we could make satisfactory progress.

That evening, December 5, we reached Piao-tsun, an extremely picturesque village of sixty families, Minchia and Lissu, which certainly had nothing very Chinese in its appearance. Situated well above the river, towards which the ground sloped away in terraced rice-fields, it owed its rustic charm to the fact that though the village was not straggling each hut had an individuality all its own. The little patch of vegetable garden in which each house stood was surrounded by a low mud wall, and further, by tall poles placed a few yards apart, up each of which twined a dense growth of runner beans. Thus the whole effect was partially to envelop the village with a wealth of green creepers from which the little thatched huts and whitewashed cottages peeped out here and there in delightful contrast to the semicircle of brown hills which rose up behind. This was our last day in the Mekong valley, though in order to complete the long stage to Shui-kin we were compelled to continue for some hours after dark.

Between the monstrous slopes of brown grass with their parks of oak and pine we now crossed deep shady gullies crowded with a much richer and more varied vegetation than anything we had seen hitherto. In one spot where the river narrowed considerably the rock-strewn slopes of this sunless gorge were clothed with jungle, where to my astonishment—for we had scarcely left one of the detestable regions of grass—the trees were wrapped with climbing Aroids and supported numerous bird's-nest ferns. Still more surprising, cascades of magnificent orchids, including a fine species of *Cymbidium*, hung in full bloom over the screes lighting up the dark forest as though by magic.

Ten minutes later we were out in the full glare of the sunshine again, and ferns, orchids, and forest had all disappeared as abruptly as they had come.

Still it was evident that we were approaching a region where, under favourable conditions, a monsoon vegetation was able to establish itself, though it was equally evident that so long as the Mekong valley consisted of a deep rent torn between high mountains, this vegetation could only exist in such small quantity as in no wise to affect the general appearance of the valley. In these deep ravines it is not because the rainfall is heavier—which it manifestly cannot be—that the vegetation is so luxurious; it is simply that here the sun never penetrates, and the heavy dews precipitated in a region of intense radiation are able to keep the vegetation moist throughout the day, whereas in the main valley the dew is sucked up as soon as the sun appears over the ridge.

As already stated we pursued our way long after sunset, the night being perfectly clear and calm. For some time we proceeded in darkness, but when at last the full moon appeared over the ridge, such a flood of brilliant light flashed into the valley that we could almost have seen to read small print. It was a wonderful sight to look down at the river winding out from the deep shadow, its black waters suddenly becoming spangled with silver as they burst over the rocks. In contrast to the open mountain-side the deep gullies which had to be crossed were dark as caves, and looking back I saw amongst the dense forest which choked them, wild bananas, their enormous leaves glistening in the brilliant moonlight. A little later and we were in Shui-kin.

CHAPTER XVII

THE LAST OF THE MEKONG

THIS was the last we were destined to see of the great Mekong river. Between the chain suspension bridge on the high road across Yunnan (lat. 25° 15′) and Samba-dhuka in Tibet (lat. 29° 35′), a distance of perhaps five hundred miles as the river flows, we had followed it altogether for seventeen days' journey, representing two hundred and fifty miles of road, and now we were about to take our final farewell. We were to turn westward from this point and cross the Mekong-Salween divide *en route* for T'eng-yueh.

I was scarcely sorry to say good-bye, for the Mekong gorge—one long ugly rent between mountains which grow more and more arid, more and more savage as we travel northwards (yet hardly improve as we travel southwards)—is an abnormality, a grim freak of nature, a thing altogether out of place.

Perhaps I had not been sufficiently ill-used by this extraordinary river to have a deep affection for it. The traveller, buffeted and bruised by storm and mountain, cherishes most the foe worthy of his steel. Nevertheless there was a strange fascination about its olive green water in winter, its boiling red floods in summer, and the ever-lasting thunder of its rapids. And its peaceful little villages, some of them hidden away in the dips between the hills, others straggling over sloping alluvial fans or perched up on some ancient river-terrace where scattered blocks of stone suggest the decay of a ruined civilisation—all these oases break the depressing monotony of naked rock and ill-nourished vegetation, delighting the eye with the beauty of their verdure and the richness of their crops.

Happy people! What do they know of the strife and turmoil of the western world? We wear ourselves out saving time in one direction that we may waste it in another, hurrying and ever hurrying through time as if we were disgusted with life, but these people think of time not in miles an hour but according to the rate at which their crops grow in the spring, and their fruits ripen in the autumn. They work that they and their families may have enough to eat and enough to wear, living and dying where they were born, where their offspring will live and die after them, as did their ancestors before them, shut in by the mountains which bar access from the outer world.

From Shui-kin we ascended the western range by one of the ridge-like spurs which stand boldly up between the deeply furrowed ravines, where the torrents thunder long and loud throughout the summer, only to die impotently away in winter. Forests of pine clothed the lower slopes, but higher up a little cultivation was carried on by a few scattered families who lived the simple life two or three thousand feet above the river. Flocks of green parrots darted, screaming shrilly, from one point to another, and once a wolf crossed the path within easy gunshot, stopped to look at me, and fled incontinently on catching sight of Ah-poh; otherwise animals and birds were as rare as men. Towards the summit the vegetation became richer and more varied, the ascent steeper, till finally, after toiling up innumerable roughly-laid steps, we reached the summit of the pass shortly after mid-day.

The Mekong, as we know, flows into the China Sea; the water which trickled down the slope in front of us was bound for the Indian Ocean, so that we stood once more at the parting of the waters. Do we realise what it means, this paltry barrier of rock separating the waters of two oceans, gradually wasting away under the assaults of the south-west monsoon? Some day when we ourselves have crossed the Great Divide, the torrents flowing to the Salween, having cut their way further back into this rugged wall, will tap the water on the other side, and the Mekong will be no more. Already the Salween flows from 1500 to 2000 feet below the level of the Mekong; already the much greater rainfall in the former valley has caused the water-parting to move so far eastwards that its crest now hangs right over the Mekong. Such a beheading of one of the great rivers of Tibet can only take place south of latitude 28°, for north of this parallel are the arid regions where the rainfall is practically the same in all the deep valleys. But here the change is surely being effected, and the tropical rainfall of the Salween valley is gradually strangling to death the ever-shrinking Mekong, flowing in its rainless gorge.

Spread out at our feet was a green valley where small villages, surrounded by cultivated fields, peeped from amongst clumps of trees; here and there shone the white walls of a temple, and in the distance Lao-wau caught the sunshine and glistened indistinctly. Beyond the valley rose another range of mountains, completely shutting out any view of the Salween, and still further away appeared the black and jagged crest of the ridge which separates the Salween from the Shweli.

On this side the country does not fall away abruptly to the Salween, as it does to the Mekong on the east, but is more cut up into hill and dale by the mountain streams which, flowing in broad valleys, run for some distance north and south before turning westwards to break through the secondary ranges so created, and join the main river. The heavy rainfall leaves its mark on the topography of the country, which is much diversified, the broad valleys being further characterised by their gently rounded form and the smoothing away of abrupt inequalities on the ridges by masses of vegetation. There were here none of those hard lines so typical of the Mekong scenery.

A steep descent, at first down well-laid steps, presently deteriorating into a narrow and slippery path through pine forests similar to those above the Mekong, brought us into the valley, where we stopped for lunch at a tiny village.

Here camellias were to be seen in flower, besides several other shrubs, and one or two betel-palms rose high above the thatched roofs, so that we were certainly approaching a region of mild winters, though immediately above us frowned the rugged cliffs of the divide we had just crossed, and it had been by no means warm up there on the pass.

Following the stream westwards through the gap it had

made in the ridge fronting us, we reached Lao-wau early in the evening and I took up my quarters in a small temple.

As previously stated I had met the Tussu of this place in the Mekong valley, but his deputy, a lame man who was carried in on the back of a friend, at once called on me to see that I had everything I wanted. A fire was set blazing on the stone floor by my bedside and gradually a crowd of some twenty men gathered round to watch me eat my supper. All—they were mostly Minchias—were friendly, and after supper I joined them round the fire while they plied me with questions, evincing the greatest interest in England. "Were there any mountains in my country?" "How far away was it?" "What crops did we grow?" "What did we eat?" "Had we a King or an Emperor?" and other questions of similar nature. Having an ejector gun, I took the opportunity of showing them what wizards the English are. "Now" I said, displaying the breech with the cartridge inside, "that cartridge will come to me when I whistle," whereupon, opening the breech till the spring was almost free, I fixed a stern eye upon it and whistled sharply, once, twice; at the same moment I surreptitiously continued till the spring was fully released, and obediently the cartridge hopped out on to my lap. This performance created roars of mirth, though for the moment the men were so obviously taken aback that they merely stared incredulously. At length, comprehending, they asked for an encore, and I responded several times for the benefit of the newcomers to our circle. I even invited the deputy chief to try his charms on the magic cartridge, but he shook his head and laughed, for alas! he could not whistle. As a matter of fact it is seldom that a Chinaman can, and when by dint of long labour he has acquired the art of producing one or two notes, he is so inordinately proud of the accomplishment that he gives one no peace.

I now varied the experiment in order to show that my miraculous power was not confined to extracting cartridges from guns, and in the same magic way opened my folding camera, by whistling—and pressing the spring at the same time. The chief having in the meanwhile practised whistling till he could produce a sibilant hiss, thought he also could open my camera in this way, and taking it from me, he held it in one hand exactly as he had seen me holding it, but with this oversight—that his finger was not on the spring, so that though he hissed at it till he was blue in the face, it resolutely refused to obey him. My gun and camera were always a source of great interest to the Chinese and to such tribes as had attained the same degree of overpowering curiosity; but the Tibetans, Lutzu, Moso, and others were comparatively indifferent to such marvels, partly, I think, owing to a far more deeply-rooted superstition, which caused them to view such altogether incomprehensible things as a camera with no little alarm.

These people were very anxious to see me fire my gun, but as there was nothing to shoot at, and as, moreover, I might have disgraced myself for ever in Lao-wau by missing it, if there had been, I did not comply with the request.

Lao-wau, with a population of about four hundred, has I believed acquired an entirely fictitious importance, for no earthly reason that I could see, for though situated on one of the few roads between the Mekong and Salween rivers, it is quite a miserable little place.

Next day we were due to reach Lu-k'ou on the Salween, but, our rather dilapidated porters being well advanced in years and possessed of only three eyes between them, we proceeded so slowly that, in spite of continuing by night, we had finally to stop some miles short of the river.

In the drier parts of the torrent bed grew tussocks of tall grass, sometimes reaching a height of twelve or fifteen feet, the big feathery inflorescences glistening like silver; this grass is extensively used for thatching the huts of the Shans and Lissus who inhabit the Salween valley. The path by which we descended the gorge, however, frequently took us through dense jungle, affording welcome shade from an uncomfortably hot sun. Evidently we were coming into a country more densely populated than any we had traversed since leaving Wei-hsi, for presently we came upon a roadside stall where pears and splendid golden oranges were temptingly displayed. From time to time we passed men and women carrying loads of cotton up to Lao-wau, and towards evening we took our meal in the camp of some muleteers who were going up the ravine with thirty or forty mules.

Plate XXXVI

The Salween ferry at Lu-k'ou

Shan village in the foot-hills of the Salween Valley near Hwei-p'o

It was already dusk when we again set forth, and there being no moon for some hours it soon grew black as ink under the trees. The path too was very bad, and the one-eyed porter, unable to judge distances accurately, was continually putting his foot over the edge of the cliff and tumbling in a heap to the ground, on one occasion at least very nearly falling over into the river. Under these circumstances progress became slower than ever, and as it was obvious that we could not possibly reach Lu-k'ou before midnight, we decided to halt at a hut perched up on the hill side in front of us, the fire-light of which shone out brightly through the open door. We reached this retreat about ten o'clock, just as the moonlight flooded into the valley, turning night into day, and found a number of men lying round the fire, for it was bitterly cold now. Two boards were soon procured, laid across two tubs, and my bed made up, but the hut, though well thatched and eminently capable of keeping out the rain, had not been built with a view to keeping out the cold. It was, indeed, more like a rude stockade than a house, the walls consisting simply of tree-trunks in the rough, planted vertically in the ground and held together by occasional cross-pieces. Consequently not only were there big gaps in the walls, but the eaves at either end from the top of the wall to the ridge-pole were entirely open, and sleeping right up against this airy partition, I awoke at an early hour half frozen.

People now began to rise in every direction, and I found that there were altogether fourteen of us asleep in the one room of that hut, our own party contributing but four. However, this fact caused no inconvenience whatever, for the hut was of ample dimensions and, as already stated, extremely well ventilated.

I was particularly impressed with the manner in which the women folk of the family, who were allotted the other end of the room, contrived their nocturnal arrangements; for they were packed into their bed, or to be more precise, under their quilt, with the skill we are accustomed to associate with the fitting of sardines into a box just too small for them. That is to say, the mother had one end of the quilt all to herself, as befitted her position, while from the other end, where one would naturally have expected her feet to be, peeped the heads of her eldest daughter, a girl of about sixteen, and two small children ; a confused lump in the middle suggested a tangle of legs, the plank bedstead being no longer than bedsteads usually are.

It fell to the lot of the eldest daughter—who was in the exasperating position of being young enough to do the house work for her mother and old enough to look after her little sisters—to arise first, and having lit the fire, swept the floor, fed the pigs, and washed herself, to get hot water for everybody else. The mother followed half an hour later, but the two small girls having, doubtless for purposes of warmth, slept naked, appeared somewhat loth to get up under the eye of a white man, even though well screened beneath their quilt. Two little heads of towzled hair peeped out from cover, and two pairs of large black eyes, round with wonder, having stared at me for some time looked at each other and laughed slyly. Presently they also dexterously slipped on a garment apiece and emerging from their end of the quilt, stood shivering in the cold.

Without waiting for a proper breakfast we set out down the ravine, and two hours' walking brought us at last to Lu-k'ou, a small village boasting a Shan Tussu and a ferry across the Salween, a combination which doubtless confers on the place more dignity than one would imagine from its insignificant appearance, though the yamen and official inn are good solid structures with tiled roofs. Even the huts, instead of being built of wood, as in the mountains, are built of mud bricks and thatched with bamboo matting or grass. Chinese influence is conspicuous here ; the people are almost entirely Shan and Lissu in origin, but Chinese in dress and speech. It is not a little curious how Chinese influence seems to segregate itself in certain places along the trade-routes, leaving the intervening country almost untouched, for south of Lu-k'ou I came across Shans living in a state of splendid savagery.

I had been in the inn only a few minutes when the Tussu himself came round to see me, but Sung having stupidly failed to inform me who the gentleman was, I very naturally mistook him for the Tussu's servant, and treated him with scant courtesy in consequence. It was the more annoying since he came purposely to invite me to the yamen, where he had at short notice prepared an excellent breakfast for me. However I soon discovered my mistake, and feeling more expansive after I had made a good meal of pork fat, cabbage, rice, and one or two other dishes, and drunk a flagon of wine, I chatted intermittently with him for an hour and took my leave with more ceremony.

The Salween is extremely narrow at Lu-k'ou, scarcely thirty yards across, flowing quietly between high sand-banks, but though the river was already fifteen or twenty feet below its summer level, the olive green water seemed to be of great depth. Immediately beyond the ferry it broadened out again, maintaining an average width of fifty or sixty yards. My aneroid registered 27·08 inches in the river bed, equivalent to an altitude of about 3000 feet.

We crossed the Salween in a big scow and presently found ourselves on a good path through the jungle, which fringed the right bank above a rock-bound shore. Strange fruits dangled above our heads, fantastic creepers wrapped themselves about the trees, and occasionally a bright flower shone out amongst the sombre vegetation of leaves. But if the right bank of the river presented a tropical aspect, it was well balanced by the left, which emphatically did not.

It was but a short distance to Pai-lou, and continuing down the valley till dusk we reached a miserable Shan hut, situated at a point where the river, divided by an island of shingle, plunged with a thunderous roar over an enormous rapid. Although the Salween valley, with its narrow belts of jungle, its innumerable rice-terraces where the buffaloes browsed lazily, its spreading *Ficus* trees standing in splendid isolation, its villages sheltered beneath palms and banana leaves, had the seal of the tropics plainly set upon it, the nights were as a matter of fact bitterly cold, for the radiation into this air, very clear towards morning after the precipitation of the heavy dew, was intense; it was indeed a climate of extremes, hot by day and cold by night, drenching rain throughout the summer, halcyon days in winter.

As night came on I was extremely thankful to sit by a roaring fire, for the cold readily penetrated the flimsy log hut, and, remembering my experience of the night before, when I was fairly stiffened in that bitter air, I slept by the fire.

The women of the establishment, who chewed betel nut continuously, happily slept in a more substantial mud hut and left us in peace. Two small boys spoke Chinese quite well and took the opportunity of asking me for some Chinese cloth, that commodity representing the height of their desires, for the moment at any rate. They were dressed in thin white cotton garments of Shan make, both old and worn, which, though all very well in the day-time, were decidedly inadequate at night, so as I happened to have some yards of stout blue Chinese cloth, and they were nice friendly little urchins, I made them a present of it, to their unbounded delight.

The cross-bow was as common here as amongst the tribes of the north, several being hung against the wall, and during our journey down the valley we daily met small children prowling about with these weapons, trying to ambush and annihilate still smaller birds, with very in-different success.

Across the river was a sheer limestone cliff, and there, plainly engraved some thirty feet above the present level, was a water mark. What a scene it must be when it rains continuously for six months on the Tibetan plateau!

At the point where the river was divided, was a curious fishing apparatus erected close inshore on the narrower branch, just above the rapid. This consisted of a small stockade, some six feet square, built of stout tree-trunks, projecting a few feet above the surface of the water, and apparently filled with boulders to stiffen them against the rush of water. It looked indeed like a solid redoubt, but it was, as a matter of fact, open under water at the upper end, enclosing a hollow space through which the river swept like a mill-race. The exit at the lower end was more contracted, and from this opening trailed a wicker basket, something like a lobster-pot, into which any fish unfortunate enough to find themselves inside the stockade were inevitably swept.

The path, still traversing a thin belt of jungle and crossing the rock-choked beds of numerous torrents, continued close beside the river; but just below the rapid was an extensive beach composed of shingle and boulders, some of the latter being of enormous size. Here then was some hint of the prodigious feats of strength performed by the Salween during the summer rise, for it was evident that these well-smoothed masses of stone were not only covered at high water, being indeed scarcely fifteen feet above the present level, but that they were also picked up and flung about in the most reckless confusion, like so much flotsam. Indeed, the rise of a big river like the Salween, which not merely receives an enormous accession of water from the melting of the snows towards its source, but is at the same time inundated throughout the length and breadth of its valley for over a thousand miles by the monsoon rains, passes belief till one has seen it. The roar of the rapid was very plainly heard in the hut where we slept, but what must it not be night and day throughout the summer when boulders weighing several tons are being ground against each other like pebbles in a rill!

The much greater volume of water brought down by the Salween was now apparent, as was also the far heavier local rainfall. The Mekong gorge may be briefly described as V-shaped in cross section, the Salween valley as U-shaped, and to this circumstance must be attributed its greater population in the monsoon area, since here there is far more land available for cultivation. The distant views we were able to obtain up or down the valley presented a charming picture of the river winding between mountainous spurs which, thrust boldly out from the dark forested ridge to the west, rose one beyond another till they became blurred and indistinct in the haze, and were thus entirely different from the burnt-up cliffs of the Mekong, standing out sharply in the immediate foreground and forbidding any view down the tortuous gorges.

On the right bank the valley rises gradually at first, the slope becoming more and more abrupt as the crest of the divide is approached. Torrents plunging down the precipices have dissected the country below into rolling foot-hills, breached by wide-mouthed openings from which project extensive alluvial fans sloping gently to the river. These contiguous fans have been built out till they now form a shelf averaging a mile in breadth and terraced throughout, stretching from the river to the base of the foot-hills from which the numerous streams debouch.   Behind, the divide rises steeply, its buttress-like ridges sweeping down in grand curves to merge imperceptibly with the undulating foot-hills.

The chief region of cultivation is this alluvial platform, though the wide valley mouths are also extensively terraced ; the chief region of occupation, the foot-hills, where in each little pocket a Shan village is tucked away.   The rugged mountain range between the Salween and Shweli is densely forested except along the immediate crest.   On the left bank, however, the dry brown mountains dip almost straight down into the river, leaving but little room for cultivation, since the torrents, far fewer in number, instead of building out wide alluvial platforms, have burst through the final range by cutting out deep gorges for themselves.   Thus no foot-hills have been formed, and it is only here and there in favoured localities that the slopes are sufficiently gentle to be terraced, villages being few in consequence.   It is evident, therefore, that though the Salween valley receives a far more copious rainfall than does the Mekong gorge, this extra precipitation is itself confined almost entirely to the Salween-Shweli divide.

The crest of this range is plainly visible from the valley, as far as the eye can reach north and south, being, indeed, but a few miles west of the river and within a day's climb ; but the summit of the Salween-Mekong divide, hanging immediately above the latter river, is not visible at all for, owing to the erratic courses of the streams flowing to the Salween on that side, a secondary ridge has, as already remarked, been blocked out in front of the main watershed.   The place of the foot-hills is really taken by the broad valley intervening between the two ridges, on the upper courses of the torrents which, in response to the distribution of the rainfall, show the reversed type of valley structure already described.   That there is only a slight rainfall on this side immediately above the Salween is amply demonstrated by the comparative poverty and scantiness of the vegetation, and still more plainly by the straight-sided narrow gorges through which the torrents enter the river.

The result of these climatic differences between the

Salween and Mekong valleys is seen in the fact that two races differing widely one from the other inhabit valleys two days' journey apart, and it is important to remember, when considering the emigrations of the tribes, that the Salween valley marks the eastern limit of jungle. It is not till we reach the true arid region in latitude 28°, when the physical conditions in both valleys become the same, that we find the people similar also. Furthermore, as a result of the unequal distribution of rainfall in the Salween valley itself, by far the larger portion of the population is confined to the right bank.

At mid-day we left the river and turned up one of the many wide-mouthed openings in the rampart of hills above. A dense undergrowth of shrubs and grasses, through which small footpaths in endless confusion led to Shan huts and villages concealed behind heavy foliage, rendered progress difficult, for the men were not well acquainted with the route. Fields of cotton and buckwheat indicated the presence of a considerable population, but so skilfully were the villages hidden that they were generally invisible from below.

A belt of uncleared scrub was commonly left round each village, and as we ascended further into the foot-hills, the grey thatched roofs became visible here and there, peeping up from amongst a tangle of betel-nut palms, bananas, huge clumps of bamboo, and tufts of tall feather-grass. The huts themselves were built entirely of bamboo, consisting simply of one or two rooms with a mud floor, the roof being quite the best part of them. The banana commonly grown here, which extends as far north as 26°, where, however, it probably does not ripen its fruits, is no doubt the dwarf species of southern China, which has been introduced with such success into some of the South Sea islands.

The women were attired in the native fashion, which was almost as flimsy as the huts, the children in no fashion at all, being naked, while the men had adopted more or less Chinese dress. It is curious to note that amongst all the tribes I came across in a state of partial absorption by the Chinese, the women seem to have been far less influenced than the men. This may be due partly to the fact that the Chinese merchants and soldiers, not having brought wives with them, have always been inclined to marry women of the tribe, at least temporarily ; hence it must not be overlooked that the men frequently are Chinamen, and as to the children, the boys would be apt to follow their fathers in the matter of dress, the girls their mothers. But no doubt it is frequently a matter of convenience amongst the men, the women alone keeping to the national costume from motives of loyalty.

Towards evening we reached a well-paved road at the head of the foot-hills, and continuing southwards, climbed up and down the endless succession of low spurs. It was an hour after dark when we eventually reached the Tussu yamen of Lien-ti, a mud house standing alone, with a few scattered huts in the immediate vicinity.

Next day, December 10, we continued southwards down the Salween valley, our road winding on over the foot-hills some distance above the river, which, however, was frequently visible through the wide gaps where the streams broke through to the shelving platform below. In every dip and hollow nestled little Shan villages, while an occasional larger village, its houses built of mud bricks, its whitewashed temple flashing in the sunshine, told a story of prosperity. Though it was the dry season, the prevailing colour of the landscape was still green, only interrupted along the terraced slopes of the little valleys, where the stubble caught the sunshine and gleamed bright yellow. At mid-day we finally left the foot-hills and began the serious ascent of the watershed by one of the ridges, halting at a small house for the night.

We were now some two thousand feet above the river, and as the sun sank down behind the towering cliffs above us, the valley was filled with a glow of such wonderful colour that no description of mine can convey any idea of its spell. While all was dark and gloomy in the depth of the valley, the setting sun caught the tops of the mountains across the river, and one forgot their bare brown slopes under the waves of crimson light which they reflected. Gradually a deep blue shadow crept up out of the valley and wrapped the hills in slumber, while a soft clinging mist seemed to precipitate itself from the atmosphere and spread over the rice-fields far below. In the gloaming the crimson died down to purple, the purple became violet, and still the glorious colours of sunset played up and down the valley. Away to the south a few wisps of cloud caught the slanting rays of the sun, which flashed like the beams of a heliograph through a gap in the black wall of rock overhead, and diffused an orange glow into the deepening blue. Then a few stars shone out, and the ridge was clearly silhouetted against the eastern sky : night had come down like a curtain. Suddenly across the valley the whole mountain side broke out in lines of rippling fire, which shot up silently out of the gloom, but it was only the dry grass and forest being fired for purposes of clearing and cultivation during the ensuing rainy season.

I have mentioned the gales of wind which throughout the summer rage up the deep valleys and tear across the passes of the dividing ranges. I think it likely that the haze noted in the dry season is due, not to moisture in the atmosphere, but to impalpable dust whirled high into the atmosphere, where it floats till brought down by next season's rain. The brilliant sunsets which we experienced every night make this the more probable, and the heavy night dews in this dry climate are perhaps due to increased radiation both by day and night owing to these fine floating particles.

The journey was marred by an incident which for a moment threatened to be serious, though happily nothing came of it. One evening, owing to a misunderstanding, I got into trouble with some Miao muleteers, and on pushing into the room at the same moment that these men half slammed the door in my face, I saw by the glow of a charcoal fire, round which they had been sitting, what appeared to be a gun barrel almost touching my chest, and in the dusk beyond, the dark outlines of three men with benches raised above their heads as though to strike me. For a moment I stood perfectly still upon the threshold taking in the situation, and then, looking the angry frightened men in the face and smiling, I stepped quietly into the room, confident that as I was plainly unarmed, no one would touch me, though I felt nervous about that long black barrel ; guns have a way of going off when least intended.

" Is this a gun ? " I said to the man behind the door, and

then, chaffing, " do you want to kill me ? "　But it was no gun after all, only a long bamboo spear.　Then the Miaos laughed, and explanations having been made, they were quickly persuaded to put down the benches and we parted good friends.　After that I was more careful than ever in my dealings with tribesmen who were unaccustomed to Europeans.

Early on the following morning we reached H'wei-po, a village consisting of four huts built on a small platform, whence we obtained a magnificent view of the Salween winding through its deep valley.　H'wei-po was apparently a place of some strategic importance, commanding one of the few roads over the Salween-Shweli watershed, for I found here a garrison of forty soldiers encamped, watching the road.　At first the officer refused quite politely to allow me to proceed, requesting me to go back to the Salween valley, but on presenting the cards Li had given me, he not only allowed me to pass, but told off two of his soldiers to escort me over the watershed.　He also gave me cards bearing directions similar in tenour to those given me by Li, which were to see me safely to T'eng-yueh, now only four days' journey distant.

Above H'wei-po the hog's-back continued in a series of ascents and descents by means of steep stairways of stone. It was slow work even with our lightly-laden porters, but with mules it would have been worse.　In some places the narrow path had been cut through rotted granite rocks, and deepened by wind and rain till the sloping banks rose high above one's head, while the path at the bottom was so narrow that one had to straddle along with a foot on either wall, a most tiring method of progression.

As we approached the summit we met the full force of a terrific blast which, in spite of the bright sunshine, chilled us through ; and away to the north we caught a glimpse of snow glittering on the Salween-Mekong watershed.　Forest had now given place to grass and thickets of dwarf bamboo, from amongst which trickled numerous streams of crystal water, all glazed with ice.　At the pass my aneroid registered 20·75 inches, so that it was several hundred feet higher than the pass by which we had crossed the Mekong-Salween divide five days previously.

The view westwards from this rugged range was grand in the extreme. Immediately below us, over the densely forested ridges and gullies, the narrow plain of the Shweli, hemmed in by precipitous slopes, stretched from north to south and faded away in the evening mists which were already rising from the valley. A ribbon of silver wriggled across the plain—it was the eastern branch of the Shweli, and hundreds of irrigated rice-fields, catching the sun already low in the heavens, gleamed irregularly in the deepening gloom.

Beyond the narrow plain rose range after range of rugged mountains, dark against the setting sun, like violet waves washing against an orange shore. Was I really back again in the Land of Deep Corrosions? Save for the misty plain below and the setting sun—a sight which, astonishing as it may seem, I had not seen for ten months—I could almost think so, for the country still maintained its formidable aspect, though the mountains were in miniature, the valleys broader; but I remembered with a thrill that behind those ridges which grew lower and lower to the south-west, the sun was setting over the golden land of Burma.

An appalling descent down steps so high that one had literally to leap from stone to stone brought us to a miserable temple, perhaps a thousand feet from the summit, and here we passed a bitterly cold night.

Next morning we rose at the first hint of dawn, the soldiers started back for their camp, and without waiting for breakfast we began the descent to the plain, which lay right at our feet. We had ascended from the Salween by a ridge, necessitating endless climbing up and down; the descent however was down a richly-forested gulley, the road so frightfully steep and slippery and so clogged with loose boulders that I could not but feel extremely thankful we had no animals with us. Yet we had as a matter of fact been on a pack-road ever since we left La-chi-mi, and Li himself had I believe brought mules over this very pass. But bearing in mind the steepness and irregularity of these long flights of stone steps, the passage of the two main mountain ranges with pack animals must be a hazardous undertaking—it is bad enough on the main road across Yunnan.

Before reaching the mouth of the gulley we crossed the spur and descended a more gentle slope to the plain, stopping at the first village we reached to take breakfast.

It was no great distance to the river, a small rapid stream, quite unnavigable and easily forded at many points, at least in winter. Dams had been constructed here and there, the water being allowed to pour through a sluice and deposit any fish which happened to be present in baskets placed for their reception, in much the same way as in the fish-traps on the Salween ; but angling with rod and line was also quite a popular pastime.

At the small market village of Lao-kai we called a second halt, and ate our meal in the street, being consequently surrounded by a crowd of curious sightseers, who were probably Chinese in everything but birth. Only the fact that many of the women did not bind their feet and were moreover tolerably good-looking suggested another element.

Pursuing our way we crossed endless rice-fields and at dusk reached the market village of Kai-t'ou, a mean and dirty little place whose inhabitants crowded round the inn door as though even the building itself had been grotesquely affected by my presence. Certainly they could not see me, for having been free from this type of curiosity throughout my travels, I found it sufficiently unbearable at the end, and hid securely in my room.

In fact the only good word I can conscientiously put in for Kai-t'ou is that, the early mornings being bitterly cold, with hard frosts, everybody was supplied with a small bamboo basket containing an earthenware pot full of red-hot charcoal, to be carried about whether one is engaged in sweeping the room or cooking the food or waiting impatiently for breakfast, as I was. This device, however, is by no means peculiar to the locality.

But if I excited curiosity, it was nothing to the furore created by the appearance of Ah-poh. Never a man passed us without remarking on his size, or the length of his hair, or his entirely unique figure, and on the following day he had the satisfaction of stampeding an entire caravan of mules, who doubtless thought he was some wild animal escaped from the jungle. After this little incident, Ah-poh

Plate XXXVII

Looking down into the Salween Valley from the Salween-Shweli
Divide near Hwei-p'o

Chain and wood bridge over the Shweli (eastern branch) near Kai-t'ou

meanwhile turning round to me with a pleased "See what I've done!" sort of expression, I thought it expedient to lead him by a rope when passing pack-mules, though even then they sidled past him in the gutter with an eye open for possibilities.

On December 13 we made a long stage, the road alternately threading the rice-fields and winding over low hills covered with pine trees, which shaded numerous graves. At the base of one of these intervening ridges masses of calcareous rock lay about in confusion, and innumerable hot springs welled up, some underfoot, others out of the tufa. Indeed, one of these springs issuing from a rock crevice formed a small geyser, jets of hot water mingled with vapour spurting out now and again to the accompaniment of a continuous gurgle inside the rock. The usual bath had been built at a point conveniently situated for the inflow of a cold stream, and here several men were washing themselves, while others sat on the rampart of stones dangling their feet in the water. Not only was hot water rising from the ground and casting up tiny fountains of sand in a score of places, but the whole marshy region was bubbling with gas, which streamed up through holes no bigger than worm-burrows, perforating the ground in all directions. Somehow the marsh with its beds of rushes, the uncouth blocks of cankered stone, heaped about in chaos, and the clouds of vapour hanging over the bathers, looked curiously out of place amongst the rice-fields which surrounded them : the sizzling rush of bubbles—it was probably marsh-gas disengaged by the action of hot water on rotten vegetation— sounded very thin and far away.

We lunched at Chiang-tso and continued traversing the same interminable rice-fields, in many of which the ploughs were already at work slopping through the deep mud behind the ponderous buffalos. Two very pretty arched wooden bridges, supported on chains in some ingenious way, span the Shweli below Chiang-tso, where the river contracts and flows gently between high wooded cliffs. Crossing by the second bridge we presently found ourselves in rolling country once more, and gradually left the eastern branch of the river behind ; we were in fact crossing the low spur which here intervenes between the two branches of the

Shweli. Not far to the south-west the T'eng-yueh volcano was sharply outlined against the glowing sky, looking very real, as though it had been recalled to life and was itself reddening the wisps of cloud with the reflection of its incandescent rocks. Here too were ancient beds of lava, the freshly ploughed fields being coloured a rich ochre from the decomposition of iron-containing compounds.

It was quite dark before we reached the scattered village of Chü-ch'ih, where we found shelter in a small house situated by the entrance to the chain suspension bridge over the western or main branch of the Shweli. Here a nice old man, who was a confirmed opium smoker but seemed hale and hearty in spite of his fifty years, made me welcome, prepared tea and a big fire, gave up his bed to me, and chatted away about T'eng-yueh and recent events.

We were off again at daylight on the following morning and again without waiting for breakfast, crossing the river by the chain bridge, which was not much shorter than the bridge spanning the combined streams lower down, on the main road; a dense white mist hung over the valley and the air was as keen as a knife-blade. After ascending for an hour we reached some little flat pockets of cultivated land wedged in amongst low wooded hills, and stopped shortly afterwards at a village for breakfast, eating our meal in the open. To say that I was excited would be putting it mildly, for we were only a dozen miles from T'eng-yueh, and leaving the men to finish their breakfast and pack at leisure, I hurried ahead.

What did it matter now that I had just tramped three hundred miles; that my hair was long and unkempt, my face pinched and bearded; that my feet were sticking out of my boots, my riding breeches torn, my coat worn through at the elbows? What did it matter that I had not changed my clothes for three weeks, nor bathed, nor combed my hair? What did anything matter! In another hour I stood at the summit of the low pass which separates the Shweli basin from the T'eng-yueh river and looked down on the charming little lake called Ch'ing-hai, surrounded by wooded hills. I was across the last watershed.

CHAPTER XVIII

BACK TO BURMA

DESCENDING the narrow tree-girt valley I soon came upon peninsulas of rice-fields thrust out toward the foot of the hills like grasping tentacles, and shortly afterwards emerged on to the T'eng-yueh plain. Prosperous-looking villages nestled close against the hills, a busy fishing population was engaged with nets and rods amongst the swamps of the river, and crowds of people were wending their way towards the city, with loads of country produce for the market. I had seen no such sight for nine months, and suddenly I felt that life on the plains also was very good indeed.

The battlemented wall of T'eng-yueh was already plainly visible and half an hour later I walked into the British Consulate, where I was welcomed by Mr C. D. Smith, the Acting-Consul. And now the astonishing news burst upon me. My caravan had not arrived, there was fighting between the revolutionists on the Tali road, and chaos in the city; the European population of five, with the exception of Mr Smith, had been compelled to go down to Bhamo soon after the outbreak, and I heard of the desperate fighting in the Yang-tze valley and of the true import of the revolution, little more than faint echoes of which had at that time penetrated to the Tibetan frontier.

The situation in T'eng-yueh was briefly this. An inconspicuous and quite incapable merchant of no social standing had successfully plotted for the murder of the three military leaders, who, on the night of October 27, were shot down and bayoneted by the soldiers; the yamens were then looted, the jail fired, and the prisoners set free. Out of all this confusion and butchery the merchant, Chiang by name, now emerged at the head of affairs, and being a man of cowardly violence he proceeded to establish his authority by murdering in the most arbitrary and brutal manner anyone who stood in his way or denounced his methods. His assurance was positively amazing, and finally he had taken it into his head that T'eng-yueh, with himself at the helm, was destined to be the independent capital of western Yunnan, and had called first upon Tali to acknowledge his authority.

But Tali, being the military centre of western Yunnan, and possessed of at least three times as many troops as T'eng-yueh could put in the field, not only rejected such a preposterous demand with the scorn it merited, but prepared an army to sally forth and smash this audacious autocrat ; upon hearing which news Chiang despatched a thousand troops to stop them.

The T'eng-yueh men marched to within two days of Tali, where they met the troops from that city, and owing, it was said, to the indiscretion of the leaders, who might well have patched things up, a fight ensued, in which the T'eng-yueh troops were worsted, a result which was only to be expected, seeing that they were for the most part raw recruits pitted against trained soldiers.

The main road to Tali being thus barred, the T'eng-yueh troops now made a détour to the south, but again encountered the defending force, and a desperate battle was fought in which the T'eng-yueh army was almost annihilated, the soldiers even fighting amongst themselves in their panic. On this occasion there were a few field guns in action, but the Tali men alone seem to have known how to handle these weapons, which they did with considerable effect.

While accepting native reports with due reservation, I have reason to think that Kin, who obtained his information almost on the spot, gave a fairly accurate account of what really took place, and he assured me that the T'eng-yueh troops with their allies from Yung-chang, two thousand men in all, had lost more than five hundred killed in three engagements with the thousand trained troops sent from Tali ! The provincial capital also, he said, hearing that Tali was in danger of succumbing to the rebels, sent eight hundred picked troops and ten guns to their succour, but by the time these reached Tali, the T'eng-yueh rebels had already been soundly thrashed. Be this as it may, the T'eng-yueh troops were recalled, and on December 22 I watched them march through the city, between four and five hundred strong, all that was left of the thousand who had gone forth to battle three weeks previously.

Meanwhile telegrams had been sent to Tali, inquiring as to the fate of my caravan, and on December 20, a welcome answer was received from the revolutionist leader saying that it was safe, and already on its way under escort.

Impatient as I was to get down to Burma, it was an interesting situation, and I walked about the city almost daily, immune from interference but by no means immune from close observation. A wave of military enthusiasm had swept over the place and even small children were to be seen playing at soldiers. On the broad city wall, inside the temple courts which had been converted into barracks, and on the downs beyond the city, recruits were drilled daily; and sentries, as slovenly in dress as in carriage, were posted with fixed bayonets in front of the chief yamens, banks, and barracks. There was not a queue to be seen in the city. Yet business went on in the market as usual, and save for the sound of the bugles, the numerous soldiers in the streets, and the revolutionist flags flapping idly above the south gate, there was nothing to suggest untoward events. Perhaps in no other country but China could such a distracting state of affairs exist with so little dislocation of business.

Chiang, the leading spirit of T'eng-yueh and the surrounding region, was simply a rebel, and as such it was impossible for anyone to treat with him. Why the army of Tali did not in turn descend upon T'eng-yueh and exact retribution from the man whose inordinate greed and ambition had stirred western Yunnan, is a mystery. He was repudiated by the revolutionist leaders of the province, and the city was for the time independent, so that no one knew what the man would do next, and though a settlement with Tali was actually arranged while I was there, it appeared likely to prove only a truce.

On December 27 my caravan arrived, and Kin reported that he had reached Tali without accident; further, however, the revolutionists forbade him to proceed, and it was not till two weeks later that the road was open again, all traffic and mails having been suspended in the meantime. However, I was greatly relieved to learn that all was well, and on December 29 we continued our journey to Bhamo.

One might reasonably have expected—certainly I myself expected—that a fortnight of civilised life in T'eng-yueh following immediately on the discomforts of a long tramp through tribal China, would prove the climax to my wanderings; that eight easy stages to Bhamo, with plenty to eat and nothing to do except read up back newspapers in a frantic endeavour to get abreast of the times, would be but a trivial interlude between disporting myself on the outskirts of civilisation at T'eng-yueh and an uninterrupted life of plenteous ease to follow. But I was premature. As a matter of fact the very comforting halt at T'eng-yueh proved but a sorry anti-climax, and having appeared unexpectedly at that city, like a strange and ragged comet out of the northern firmament into the light of day, I burst still more noisily on my friends in Bhamo—an entirely unrehearsed effect.

The immediate cause of my discomfiture was Ah-poh, who having curled his tail tightly over his back and made incomprehensible overtures to every dog he met without obvious success, disappeared entirely before we had proceeded two miles. We spent some time scouring the hollows for him, but since it was only too evident he had been fatally attracted by the glories of city life, it was no use wasting further time, and telling Kin to go on ahead and catch up the mules, I rode back to the Consulate to report the lugubrious news. My mules I did not see again till they arrived in Bhamo, eight days later; Kin I saw for half-an-hour on the third morning, under circumstances about to be recorded.

By the time I set out again it was nearly mid-day, and travelling as fast as I could over the hills and across the stiffened lava beds of the old T'eng-yueh volcano which loomed up to the north, I mistook the road just as I had done ten months previously in another direction, and crossed the river. It was an hour after dark when I eventually rode into Nantien, the usual stage, fully expecting to find Kin standing at the inn door waiting for me with the plaintive remark that hot water and supper were ready, and my bed made; it was therefore rather distressing to find nothing at all, neither men nor mules.

There would have been no great hardship in this, for I could speak Chinese with sufficient fluency to get all I wanted, or if not, to take it, and I was thoroughly hardened to Chinese food and the casual ward for the night. But unfortunately, as already stated, I had gratuitously crossed the river by the bridge, and in order to reach Nantien found it necessary to recross it some miles lower down where there was no bridge, but only such fords as heaven vouchsafed. Consequently I was both wet and cold, and the disagreeable prospect of having to sleep in my wet clothes did not appeal to me at all.

One needs to wade and swim a pony across a deep and swiftly flowing river to get an idea of its possibilities for raising a conflict of emotions. Beauty after some preliminary hesitation having ventured in, my own sensation was one of complete bewilderment for several minutes, and I found it quite impossible, once the water was up to the pony's middle, to resolve the different motions. The water was spinning past in one direction, Beauty was struggling diagonally across in another, and at the same time being washed down stream, and the net result was that my head whirled round till I lost my bearings completely and nearly fell out of the saddle. Though already some distance below the landing place for which I had originally headed him, Beauty still kept his feet, and we were scarcely ten yards from the bank when he suddenly went down with a plopping splash; we had floundered into a deep channel and he was swimming.

My first conscious impulse was emphatically to jump off, but happily some higher instinct, bred perhaps of familiarity with similar situations, declared itself, and I found myself gripping the saddle more tightly than before. The bank was high, and Beauty, though he had but a few yards to swim, was being washed down stream at a great pace.

Would he miss the only landing place? No! making a brave effort, his feet touched bottom, and he struggled up the bank.

It may easily be imagined then that I was thoroughly wet through half-way above my knees, and prospects for the night jeopardised in consequence. Moreover the wretched inn people, instead of having a blazing fire on the floor in the middle of the room, had practically no fire at all, and that skilfully hidden within the depths of a mud cooking range, and my powers of persuasion were heavily taxed before they would consent to go out and buy damp wood to make me a special smoky fire of my own; nor did they show that enthusiasm to dry my wet clothes that the Tibetans would have displayed. However, I had supper with some young blades who were going down to Bhamo, borrowed a pair of Chinese trousers, begged a quilt, and rolling myself up, slept with more success than I could have hoped for considering that I lay on a straw pallet. Next morning it was necessary to decide on a plan of action in view of the possibility that the mules, having waited for me on the previous day, were still some distance behind; wherefore advance rather than retreat was evidently my only plan, and my clothes having been dried in the meantime, I started after breakfast on the second stage of the journey, reaching Kan-ngai at sunset. Evidently the mules were not in front of me now.

When I came through Kan-ngai early in March the people had absolutely refused to take me in—the Pien-ma incident was apparently rankling—and it will be remembered I had sought shelter in the village schoolhouse. I was now for the second time refused admittance at two consecutive inns, and in no honeyed phrases either, the excuses given being calculated to annoy rather than to propitiate. Evidently Europeans were not in demand in this village. I have remarked the same independence more than once in different parts of China—an insignificant village in a region outwardly friendly will preserve *en bloc* an attitude of bitter hostility towards the European for no apparent reason. Such a thing is scarcely to be accounted for on the assumption that a single European has behaved indiscreetly in the past for, unless he had made matters wonderfully warm, it is almost inconceivable that an entire village would exhibit sufficient public spirit to espouse the cause of the victims in a campaign of retaliation against all Europeans upon all future occasions. It is a provincialism for which I have never been able to account satisfactorily, so I simply ascribe the phenomenon to local peculiarities of temperament; and there is no doubt that this malady is prevalent at Kan-ngai in a conspicuously virulent form.

Scorning to wander from inn to inn seeking lodgings for myself and pony, I shook the dust of the wretched place from my feet, and went on some six or eight miles to a little Shan village, screened behind groves of bamboo. It was just dark when I turned aside from the high road.

Crossing the ditch we squeezed through the narrow gateway in the mud wall and I at once found myself amongst bamboo huts thatched with straw. Near at hand a gate stood temptingly open, and entering the compound I called out to the occupants of the hut, where a light burned brightly. The Shan who appeared in answer to my summons—a prematurely aged and skinny figure much disfigured about the mouth from chewing betel nut—happily had at his command at least a smattering of Chinese, but though willing, he had no accommodation to offer me and I persuaded him accordingly to guide me elsewhere. This he did, and though his friends at first refused to open the gate, even at his recommendation—at least I supposed he was recommending me to their care as a harmless and possibly remunerative guest, in spite of the bellicose attitude of two dogs—the actual production of an Indian rupee acted like magic; a few more words were exchanged, and two men came out to the wicket, which was immediately opened. Within five minutes, Beauty's wants having been attended to in the meantime, I was sitting in the hut surrounded by half-a-dozen friendly people, while a pretty Shan girl set about preparing me a meal of rice, eggs, and vegetables.

I had also a small flitch of bacon which I had forcibly annexed at lunch time, owing to the sordid behaviour of an old woman, who having stated the price of my modest repast and received an Indian rupee by way of payment, thereupon outrageously violated the rate of exchange.

Nothing would induce her to give me the right change, but redress was at hand, and to square the account I picked up two eggs and a lump of bacon which were lying on the table and put them in my pocket. As I departed the woman protested feebly that the foreigner was walking off with the eggs and bacon, but as neither she nor anyone else seemed in the least surprised or showed any signs of hostility, I felt quite justified in my violent purchase.

The very next day I was again victimised by a Shan woman, who kept a roadside stall at which I elected to take a little nourishment; and this offence was the more heinous because she had just refused to serve me at all unless I first gave her the money, a proposition which I scouted indignantly. As a counter-move she advanced the price of her goods a hundred per cent.!

After I had eaten, we all—namely three men, three women, several children and myself—sat round a small fire of burning straw, and I was requested to tell the news from T'eng-yueh, my story being subsequently translated into Shan for the benefit of the women and younger men. Finally a bed was made up for me on a little platform just beneath the roof of the buffalo shed, and being weary I retired. The bed consisted of a quilt and a very stiff, but clean cotton sheet laid on the straw, and but for the fact that the buffalos immediately underneath nearly brought the whole shed down in the middle of the night, when they scratched themselves against the posts, and the presence of a fowl perched just above my head, I should have slept well. It did not matter that the bedroom thus hastily prepared for the guest was open on three sides, because the huts themselves being made of bamboo plastered with mud, which had for the most part dropped away, were so full of holes and crevices that they were almost equally well ventilated.

The women of the establishment rose about three o'clock, while it was still of course quite dark, to pound rice, and as the fowl and the buffalos were making themselves prominent at the same time, I myself got up with the first hint of daylight and asked for breakfast, which however was a lamentable time in making its appearance. Meanwhile I passed the time trying to purchase silver bracelets from the women. These bracelets, almost the only ornaments worn by the Shans here, are about a quarter of an inch in thickness, but the ends are not quite joined up, and I was disappointed to find on examining them that they were not, as I had supposed, made of solid silver; consequently when my friends demanded ten rupees for a single bracelet, I found it impossible to trade.

As stated, the Shan women in this region affect little personal adornment, doubtless having been subdued in that respect by contact with the stolid Chinese. They are sufficiently characterised by their chimney-pot turbans and tightly-wound puttees; the whole attire, skirt, jacket and all being of dark blue or black cotton cloth, relieved only by the coarse silver bangles referred to. On festive occasions, however, they seem to revert to type, for in one village I passed through there were some of the component parts of a wedding procession—bridesmaids maybe—decked in gay colours and much jewellery, ear-rings, bracelets, silver hoops round the neck, and silver plates set in the hair; all bearing a distinct resemblance to the outfit of the Moso bride.

The men, however, have adopted Chinese dress almost entirely, except when the Swabwas desire to create an impression in the city, where a chief will sometimes appear in a most ludicrous parody of European styles, a frock coat supplemented by a bowler hat, for instance. A Swabwa in full native dress, however, is a gorgeous sight.

How different again are the Kachins, occasionally met with on this road and fairly numerous in Bhamo itself, the dirtiest and ugliest tribe I have ever come across, but with a fine taste in dress! Ugly—yet in the dim candle-lit hall of the Aracan pagoda at Mandalay I saw, bowed meekly in prayer before the golden Buddha, such an innocent-looking vision of Kachin beauty as will not be forgotten.

Their most obvious peculiarity—I speak of the women—is the girdle of rattan, like fifty feet or so of telegraph wire coiled round the waist and again in lesser amount round the legs just below the knee, to improve their walking powers, they say. The men carry beautifully wrought cotton bags, fringed, and set with silver bells, beads, and inlaid work, in shape similar to those carried by the Lutzu and other tribes, but of far more skilful workmanship.

While I was engaged in bargaining with my friends for bracelets, each of us in a different language which however did not hinder negotiations in the least, who should suddenly walk into the compound but my faithful man Kin!

He related how after the dog incident he had overtaken the mules during their mid-day halt, which had been unduly prolonged on my account, and how, seeing no signs of me, the caravan had continued on its way till nightfall, eventually stopping some miles short of Nantien, where they arrived at twelve o'clock on the second day, three hours after my departure. Here however they had news of me, and arranging for the mules to follow as quickly as possible, Kin snatched up a few miscellaneous articles with which to succour me in case of necessity and hurried ahead, reaching Kan-ngai the same evening, where he obtained further news of me, and would indeed have found me had I not been denied admission to the inns. Leaving early in the morning of the third day, he had found me as I have related.

I had felt rather vexed at the mishap, since I had anticipated a perfectly quiet journey to Bhamo, surrounded by the comforts and conveniences which I possessed. Yet, such is the perverseness of human nature, now that my caravan was within easy reach, I had no desire to rejoin it, and deliberately elected to continue the journey as it had begun. There were reasons of course. In the first place, Kin was not certain where the caravan was, but it was at least half a stage behind, and waiting for it was likely to prove irksome; secondly I had, under the ban of adversity, reconstructed my plans, and did not feel inclined to alter them again now that I had adapted myself to the conditions of a rapid journey to Bhamo. A little bedding would have been a comfort certainly, but to await the advent of the mules was to waste half the day, so I horrified Kin by announcing my intention of going straight on; and having arranged where to meet him in Bhamo, I gave the friendly Shans solid proof of my appreciation for their kindness, and rode away. That day we covered two stages without much effort, for the road was good and the distances not great. Towards evening I shot two pigeons for the pot, and plucking them as I rode along, gave them to the innkeeper at Manshien bidding him cook them for supper. It was New Year's Eve, my last night in China.

Starting again before the sun was in the valley on New Year's Day, I crossed the frontier and entered the forests of Burma. In the day time there were gibbons to be seen dropping heavily with great splashing of leaves from tree to tree, and flocks of myna birds performing droll antics; by night, the occasional sparkle of a fire-fly and that curious all-pervading sound which forms a kind of background to the tropic night, the drone of cicads. But it was by night especially that the roar of the Taping river in the gorge below carried me in fancy away back to the vast solitudes of the Land of Deep Corrosions.

I reached Kolongkha at eight o'clock at night after a ride of forty miles, only to find that the fundamental resources of a dâk bungalow are far inferior to those of the meanest Chinese inn; there were indeed crockery and glassware and excellent furniture, all at the disposal of the traveller, but even a hungry man cannot eat crockery, and if it is a choice between bedsteads and blankets the latter will be found the most generally useful. I had arrived so late that it was impossible to get anything to eat, and so cold did it become towards morning that I was eventually driven to pull down the door curtains and wrap myself in them.

Starting at 5 a.m. without any breakfast, since there was no guarantee that I should ever get any, however long the delay, I reached Momawk at two o'clock, and finding a Shan establishment where they catered for wayfarers, fared sumptuously on biscuits and coffee. Another nine miles down the dusty high road, and I was back in Bhamo, just over ten months since my departure. My travels were over.

.   .   .   .

During five days spent in Bhamo awaiting the caravan, which arrived safely on January 5, I stayed with my friends Mr E. B. Howell and Mr Joly, who were still exiled from T'eng-yueh on account of the complete disorganisation of the Chinese Customs service, and on January 8, I left for Mandalay by river. Four days later I was in Rangoon, the guest of a very old friend, now Professor of Physics in Government College, till the boat sailed on January 25 for England and home.

CHAPTER XIX

THE LAND OF DEEP CORROSIONS[1]

In the following chapter I have collected together some of the botanical and geological facts mentioned in the preceding pages and have attempted to draw certain conclusions from them respecting the geological history of the strange country I have been describing.

Stand on any one of the high passes which notch the Yang-tze-Mekong watershed—little crenellations they appear from below in this great rock-wall, buttressed by tremendous towers of limestone and crested with jagged spires—then, with the dawn behind and a clear winter sky overhead, look out over the wilderness.

Straight across the deep gulf in front, so near that it seems almost within hail, a mountainous ridge rises up from invisible depths below, its barren slopes, scorched and shrivelled by a wind as from hell's mouth, flashing in the sunshine, but crowned above by dark green clustering forests growing thinner and thinner as the dwarfed trees struggle up towards the foot of the screes. Behind it, slashed from base to summit by dark ravines which separate spur from spur, rises another ridge; and beyond that another, and another—range beyond range peeping up out of the west to grow dimmer and bluer till earth and heaven meet, while to north and south they fade away and finally melt into the infinite distance; here and there sombre forest and shining scree being broken by a pyramid of snow glittering in the morning sunlight.

How near they look, these clear-cut ridges huddled together like waves crowding up the shore out of a rough

[1] I have used 'corrosion' in preference to the more obvious term 'erosion' in accordance with the latest definitions. 'Corrosion' is now limited to the vertical digging work of rivers or glaciers, and implies wear and tear by mechanically transported material, by which means gorges are formed (*vide* Paper by Prof. J. W. Gregory, F.R.S. in the *Geographical Journal*, February 1911).

sea! Yet the black gashes between them recall to us those grim gorges where the rivers foam and thunder, flinging themselves irresistibly against iron-bound cliffs, swinging round towering headlands, cutting, grinding, pounding their way southwards. But of all this din not a murmur floats up to us here; not even the little torrent which has its source at the foot of the pass can be heard. All is silent, immense.

Such is the scene, and yet there is spread out before the traveller not so much a land of high mountains as a land of deep valleys; it is not this barrier beyond barrier, peak on peak which he sees in splendid array that impresses him, but these deep gloomy gorges into which none but the eagles wheeling far overhead can peer; gorges whose presence is realised rather than seen, with black shadows torn into every spur.

This same idea of their country has been gained by the Tibetans themselves, for Mr Edgar tells me that a classical name for Tibet is *Ngam-grog-chi* which he freely translates 'The Land of Deep Corrosions'; and for south-eastern Tibet at least no more appropriate title could be devised.

How near to each other these rivers flow may be gathered from a consideration of the time it takes to cross the high ridges intervening. Thus Mr Edgar, starting from Menkong crossed the Salween, the Mekong at Y'a-k'a-lo, and the Yang-tze below Batang within the week. And when we consider that these are three of the biggest rivers in Asia, one of them flowing to the Indian Ocean, the others to the two extremities of the China Sea, we can dimly realise something of the extraordinary nature of the country.

From Tsu-kou on the Mekong over the Sie-la to the Lutzu villages on the Salween can be accomplished by lightly loaded porters in three days, and the natives themselves frequently do the journey in two; similarly it would not be difficult to cross from Londre, just above the Mekong, to the Salween *via* the Chung-tsung-la in two days, or by the Doker-la in three. On the other hand, crossing from the Salween to the Mekong takes rather longer, at least in my experience, probably because the watershed, instead of being symmetrically placed between the two rivers, is much nearer the Mekong; moreover the tremendous rains in the Salween valley south of the arid region have dissected the mountains on that side to a far greater extent, and as the torrents nearly always flow southwards parallel to the main river for some distance before their junction, a tiresome succession of deep valleys and high spurs needs to be crossed; this however is not the case with the Doker-la and the passes of the arid region to the north.

From the Mekong to the Yang-tze is a journey which can easily be accomplished in four days over any of the passes I crossed, though personally I always set out from A-tun-tsi, which stands at a considerably higher elevation, thus requiring only three days for the journey. Following small roads by a more direct route, one could undoubtedly cross from river to river in three days without much difficulty.

Here, too, I found the journey longer when made from the Yang-tze side than when made from the Mekong side, and for exactly the same reason, namely that south of the arid region the Yang-tze valley receives a greater rainfall than does the Mekong valley, though a much smaller rainfall than the Salween valley.

This is curious. Evidently the enclosing cliffs of the Mekong gorge—a mere rift in the rocks, its basin for three degrees of latitude not exceeding fifty miles in extreme breadth—are of insufficient height and extent to check the rain-bearing winds from the west, which, having drenched the mountains as far east as the Salween, pass right over the Mekong basin and throw down the remainder of their moisture on the higher peaks towards the Yang-tze. How thoroughly this rift-like character of the Mekong is maintained is well illustrated by the passage of the watershed to the west in latitude 26°. Ascending from the river, we reached the crest of the watershed in five hours without any effort, the porters, who were carrying moderately heavy loads, taking things quite easily. But, from there, the descent to the Salween occupied a day and a half. Such is the great divide which separates the water flowing to the Indian Ocean from that going into the China Sea.

This cramming of the western watershed against the Mekong is no doubt largely a result of the heavier rainfall in the Salween valley, causing the streams which flow down that side to cut their way back at the head much more rapidly; indeed, the comparative volumes of the streams flowing to the Salween and Mekong respectively make this abundantly evident. Other things remaining unchanged, the final result of this must be to divert the Mekong itself at some point south of the arid region, beyond which this great difference of precipitation is not found (latitude 29°), and cause it to flow into the Salween, especially as it already flows at a very considerable elevation above that river.

The Mekong valley has the further peculiarity already alluded to, that a rainy belt is interpolated between the arid region north of Yang-tsa and what may be called the dry region, scarcely so intense as the former, of the middle Mekong. This rainy belt, while not comparable in richness of vegetation to the Salween forests, nevertheless receives a heavy summer rainfall, due perhaps to peculiar local conditions.

In the present chapter I propose to bring forward a few facts which may help to give some idea of what tremendous forces have been moulding this land into its present fantastic form, and at the same time to hazard a few suggestions as to the means by which this may have been accomplished.

Viewed as a whole, the region seems to have been subjected to terrific lateral pressure, either acting simultaneously from east and west, or more probably from one side only, the other side being crushed against an unyielding barrier which, by preventing any actual movement of the mass so caught, has compelled it to ruckle up in parallel ridges as one might ruckle up a piece of cloth.

This, indeed, is the appearance which the country presents, and a glance at the map of Asia will suggest a cause for this squeezing, namely the proximity of the two great mountain ranges of Tibet, the Himalaya and Trans-Himalaya, whose axes of uplift run east and west; and it is not impossible that the uplift of these stupendous ridges was accompanied by a slight lateral motion eastwards, jamming a comparatively narrow tract of country against an unyielding mass in Western China, and thus forcing it to occupy less space, which of course it could only do by throwing itself into folds at right angles to the direction from which the pressure was applied. It is important to note that there stretches eastwards from the Tibetan plateau a mountain range which reaches substantially across China, forming the watershed between the Yellow River and the Yang-tze basins; and in fact all the greatest mountain ranges of Asia trend east and west. It is only this small area in south-eastern Tibet that has for some reason resisted the movement, with the results seen.

Assuming such an origin for these great parallel ridges there are good reasons for believing that the lateral movement came from the west, not from the east, a subject to which we must refer presently.

The Tsanpo river, flowing eastwards in the deep valley between the Himalayan and Trans-Himalayan ranges, came up against the most westerly of the mountain barriers thus raised, and swung away to the south and west; the Yang-tze, rising to the east of the last big ridge, followed the trend of the mountains till a region was reached where the effect of the squeezing was inappreciable, and, no longer fettered by local conditions, turned away to the east under the predominant influence of the main uplift. But the Irrawaddy, Salween, and Mekong, being caught right in the midst of the pinched area maintained their southward direction throughout their courses.

In far western Ssu-chuan the main ridges still run approximately north and south, but as the Yang-tze has by this time made its great eastern bend, the rivers coming down from the north drain into it. Were this not the case, however—did the Yang-tze like the Mekong continue southwards—these rivers themselves would, I think, have turned eastwards independently under the final influence of the main uplift. Far western Ssu-chuan seems to represent the unyielding barrier referred to above, against which the Tibetan ranges pushed in vain, and as such it has been crumpled up in some confusion, two sets of movements having been superimposed one upon the other. The effect of the lateral movement is still sufficiently pronounced to determine the courses of the rivers from north to south, but not sufficient to obliterate completely the main axes of uplift running east and west.

This is hypothesis. To substantiate such a theory the

most fundamental necessity would be to show that our parallel ridges were of post-Tertiary age, since the Himalaya were uplifted during Tertiary times; but it is impossible to determine what earth-movements have taken place in a country of which not even the roughest geological maps are available. Nevertheless, such an hypothesis may not be altogether useless even when based simply on impressions derived from looking broadly at the country from several points of view.

I do not believe that the region of the parallel rivers simply represents a line of weakness along which the waters of the Tibetan plateau have found a convenient outlet in this direction. Terrific as is the amount of spadework which they have undoubtedly performed, I doubt if the life of any single river is sufficiently long for the performance of such a prodigious task. There are, in fact, reasons for believing that such a piece of work was never accomplished, and more reasons for thinking that dynamic forces, whether in the form of the lateral motion suggested, or as direct forces of upheaval, have been at work.

The rocks show plainly enough that the entire region has been pounded in every direction, and though the symptoms of volcanic activity are those associated with the final phases of vulcanism, not such as are typical of its inception, they yet remain to tell the story of probable cataclysms in the past.

Hot springs are to be found issuing from the bases of every range. I have seen them in the valleys of the Yang-tze, Mekong, and Salween, further south in the valley of the Shweli, and continuing westwards, within the basin of the Irrawaddy. Perceptible earthquake shocks occur from time to time at A-tun-tsi and Batang, and no doubt a seismograph would be continually recording slight tremors; the numerous landslips which take place in the A-tun-tsi district, for instance, suggest a certain amount of instability not wholly to be accounted for by the heavy rains.

A little to the north of T'eng-yueh, on the low water-shed between the Irrawaddy and Shweli river-systems, an ancient volcano stands sentinel over the remnants of its lava beds, which have flowed down beyond T'eng-yueh on the one hand and across the Shweli valley on the other.

The cone and crater still preserve their form admirably, and Mr Coggin-Brown of the Indian Geological Survey, who has examined the country, considers that the volcano was active within the last four hundred years, perhaps at the time that western Ssu-chuan was so badly shaken and Batang engulfed.

The whole of Western China, for a long distance east and north of the region we are here concerned with, shows similar evidences of waning vulcanism, and throughout their length these parallel ridges exhibit unmistakable signs of volcanic activity with such persistent regularity that it is fair to assume they are situated on a line of weakness in the earth's crust which passes through Java, up the backbone of the Malay Peninsula, and thence by way of Burma and the Shan States northwards through the region indicated as far as the plateau of Tibet. The northern half of this great fissure has had its day, and the centre of volcanic activity has shifted southwards, so that there is reason for supposing that a wave of vulcanism has passed down from High Asia, and indeed the configuration of the land, running out into a long narrow peninsula which finally tails off into a chain of islands, suggests a gradual convergence of dynamic forces from north to south.

These evidences of volcanic activity, while by no means proving that the ridges have been separately heaved up into position, or that a semi-plastic crust has been squashed between two irresistible forces, make one or other explanation plausible. On the upper Mekong at least the direction of the river's flow bears no relation to the dip of the rocks in the almost continuous series of gorges through which it fights its way, for I have frequently observed in these gorges that the strata, tilted nearly vertical, dip in a direction either at right angles to that in which the river flows, or parallel to it, and the torrents also may cleave their way through to the main gorge between vertically-tilted slabs in one place, and across their sawn edges in another.

But without assuming that any one of these rivers is alone responsible for carving out its valley, evidence is not lacking that they have scoured their beds to a considerable depth—indeed it could not be otherwise. In a limestone gorge on the Mekong I noticed across the river the remains of pot-holes and smoothed cavities in the rock between low and high water-marks, and above the road exactly similar pot-holes were visible thirty or forty feet higher up, well above the highest flood-mark. Again, at Samba-dhuka in Tibet, the confined waters of the Mekong race between fluted limestone cliffs through which the river has evidently cut its way, and a short distance back from the top of the cliff, which is several tens of feet above the highest level now reached by the water, a second fluted wall of limestone caps a small river-terrace, and may be traced for some hundreds of yards. So exactly similar is it to that which now confines the Mekong, that I could not doubt they owed their origin to the same cause.

For let us remember the summer rise in these confined gorges—due to the terrific rainfall on the Tibetan plateau being synchronous with the melting of the snows—is very great. On the upper Salween in December, when the water had not yet touched its lowest level, I noticed a water-mark on the face of a gorge nearly thirty feet higher up, and this some distance south of that country to which the above descriptive name is more particularly given. A rise of thirty feet in a river averaging sixty yards in breadth and flowing with a strong current implies a force which is almost beyond belief till one has seen it at work. We have already seen how in roughly the same latitude the Yang-tze, Mekong, and Salween flow at successively lower altitudes, and still further south this difference appears to be accentuated in the case of the two latter rivers, for Major Davies states that on the main T'eng-yueh-Tali road (lat. 25°) the height of the Mekong above the Salween is as much as 1700 feet.

The rapid changes of climate—or I should say perhaps its local character, since it is only the traveller who passes swiftly from a region of continuous and appalling rains to one of extreme desiccation—I have attempted to describe in my travels; but as the subject is one of great importance not only to the present aspect of the country but also to the changes which may be wrought here in the future of geological time, it will be necessary to discuss the subject in more general terms.

The prevailing wind is undoubtedly the south-west monsoon which, blowing across the plains of Assam and beating against these mountainous barriers, deluges them one after the other till the Salween-Mekong divide is reached. And then comes a change.

North of Lat. 28° this ridge rises abruptly from a moderate elevation to the stupendous peaks of K'a-gur-pu, and thence continues northwards into Tibet to the conspicuous snowy range of Ta-miu; and the important point is that, from K'a-gur-pu northwards, the height of this range becomes a vital factor in the distribution of climate.

In like manner, both the Salween-Irrawaddy divide to the west, as I observed from above the Salween, and the Mekong-Yang-tze divide to the east, which I crossed by three passes, receive a sudden considerable uplift in about the same latitude.

The effect of this is extraordinary. The Salween-Mekong watershed itself, on account of the overwhelming height of K'a-gur-pu, still receives a very big rainfall for some distance north of the rain-screen at any rate, but by the time the winds have crossed this great range, they have been robbed of nearly all their moisture, and the Mekong-Yang-tze divide, instead of being clothed with dense forests and waving meadows of alpine flowers, presents vast stretches of barren scree, towering pillars of naked limestone, grim rocky ridges, and an aspect so drear and bleak that the scenery appals one. The snow-line stands at an enormous elevation—little less than 19,000 feet I think, and consequently the passes, high though they be, are open most of the year.

The passes over the Salween-Mekong divide are considerably lower, but owing to the far heavier rainfall which this receives in the neighbourhood of K'a-gur-pu, are open no longer than those on the next watershed to the east. The Sie-la above Tsu-kou for example, which is about 14,000 feet, was under deep snow when we crossed it in June, and was almost impassable when we crossed during the second week of November. It could hardly have been clear before the end of June, by which time the Doker-la somewhat further north was just clear, and snow was certainly falling there again in October. The main pass above Londre, situated between the two passes mentioned above— the Chun-tsung-la as it is called—is barely 13,000 feet high,

Plate XXXVIII

The Yang-tze in the arid region, looking south from near Batang

The Salween in the arid region, below La-kor-ah

but deep snow lay as low as 12,000 feet when we crossed on November 5, while at the same time there was no trace of snow on the Mekong-Yang-tze divide below 15,000 feet. These facts lead me to think that the snow-line is considerably lower than 19,000 feet on the Salween-Mekong watershed and the ridges to the west of it.

Further evidence for the sudden cessation of the monsoon rainfall on the Mekong-Yang-tze ridge is afforded by a glance at the vegetation on the one hand and at the structure of the peaks on the other. To the former I have already alluded, comparing the dense forests of deciduous-leaved trees and the meadows of tall grasses and magnificent flowers, with the sombre forests of conifers, the barren screes, and the alpine pastures clothed with dwarf flowers. Two great square buttresses of roughly-hewn rock rise above the snow-line on the Mekong-Yang-tze divide; the snow clings to their wall-like faces in patches and bands following the lines of stratification, and the glaciers descend from them like cataracts of ice. They have been carved out of these ridges by dry denuding agents, by sunshine and frost, which have splintered them this way and that. The watersheds to the west, on the other hand, are crowned by gently-rounded peaks and graceful pyramids tapering up into needle-like summits, which rise into the regions of eternal snow and frequently remind us of the majestic Matterhorn.

In the Salween valley the heavy summer rainfall continues as far north as T'sam-p'u-t'ong, in spite of the snow-clad range overlooking that village. I found rich forest in the shady ravines, and the epiphytic orchids and ferns, numerous lianas, and incipient plank-buttresses supporting the tall straight-limbed trees gave a distinct hint of the tropics; and this exuberance of vegetation is continued southwards throughout the Salween valley.

In the corresponding region of the Mekong, however, conditions are quite different. I followed this valley southwards almost continuously for 200 miles as the river flows and was astonished at its barren aspect. Forest, or more generally scrub, predominates in the rainy belt between Tsu-kou and Hsiao-wei-hsi, but south of this the valley becomes arid once more, and only grass partially conceals

the scorched rocks.   True, in the deep shady gullies the
vegetation becomes rich and varied with large trees, ferns,
orchids and creepers, even the wild banana being quite
common in latitude 26°, but these oases bear but a poor
relation to the whole valley, which can only be described
as barren and shrivelled up in appearance, terribly wearisome
to traverse day after day.

As to the valleys west of the Salween, I met in A-tun-
tsi two very interesting Chinese travellers who were able to
shed some light on the matter.   One of them had been from
the Mekong to India with Prince Henri d'Orléans in 1895,
the other had only two years previously made a journey, for
political purposes one must suppose, from the Mekong as
far west as the 'Nmai-kha, some distance to the north of
Prince Henri's route.   Both of them described the country
as covered with dense jungle, the summer rain as incessant,
and spoke lugubriously of mosquitos, leeches, tigers, and
other pests associated with such a climate.   It is therefore
abundantly evident that the region west of the Salween gets
the full benefit of the south-west monsoon.

It is not till we get further north, where all three ridges
are suddenly elevated to a great altitude, that the effect of
the monsoon begins to wear off.

A little above T'sam-p'u-t'ong on the Salween the forests
suddenly disappear; gone are the ferns and orchids, the
creepers and all the great wealth of vegetation; and gone
too is that incessant pitiless summer rain.

Suddenly the rocks begin to rise steeply from the river,
which becomes more and more shut in by gigantic cliffs,
apparently bare, but actually supporting a scanty vegetation
of succulent herbs, withered cryptogams, and dwarf shrubs.
The valley becomes still more confined, the mountains
grow steeper, gorge follows gorge.   Immense screes devoid
of any vestiges of life tower upwards for many hundreds
of feet.   A scorching wind rages up the valley sucking the
vitality from every living thing, as though anxious to reduce
all to the condition of these naked rocks.   The sun glares
down out of a blue sky intolerable in its monotony, and
is reflected from the light-coloured cliffs and screes seen
distorted through a quivering atmosphere.   Yet on the
mountainous ridges to east and west is ceaseless rain!

With the abrupt change of climate the people change too. On the Mekong north of Yang-tsa the same conditions are repeated, if possible with even greater intensity. Never have I seen cliffs so stark, so hideously bare, gorges so forbidding. The river thunders wildly down its deep gutter, draining the Roof of the World—'an exaggerated mountain torrent,' as the French priest Degardine aptly terms it. Yet here and there, as on the Salween, a mountain torrent dashing down has thrust out a small alluvial fan hanging far up above the river, where cultivation is possible. It is indeed a beautiful sight to see one of these oases in the spring, green with wheat and walnut trees, in the summer golden with the ripening corn, where are concealed the pretty Tibetan 'manor' houses scattered down the terraced slope, while all around the enclosing cliffs rise bare and drear.

It is the same on the Yang-tze flowing behind its rocky barrier to the east, but here the little villages, in some places perched far up on the steep valley walls, in others nestling down close by the water, are more numerous, so much bigger is the Yang-tze. Man has wrestled rudely with the problems of life by the River of Golden Sand for untold centuries, the mere thought of which, carrying one back to the ages when the river was perhaps young, makes the head reel.

Thus all three rivers flow through an arid region, the southern limit of which may be set down as 28° 10′, while northwards it extends into the unexplored regions of Tibet. But whereas the Mekong flows through arid gorges for a long distance south of this, and the Yang-tze valley has, I imagine, a rainfall comparable to that of England, the Salween alone exhibits the two extremes of climate separated by an absolutely sharp line, the position of which could be marked down within a mile either way; and in this respect at least it is the most marvellous river of all. It is not difficult to account for the arid region, however, bearing in mind the arrangement of the mountains and the direction of the rain-bearing wind.

If the Salween-Mekong divide so thoroughly exhausts the winds of their moisture that the next ridge over which they sweep presents in consequence such different scenic features and such a diverse flora as that we have seen on the Mekong-Yangtze divide, much more will this be the case in the deep gutters which separate ridge from ridge, since it is here that the effect of the rain-screen makes itself felt in its fullest intensity.

The positions of the great snowy ranges correspond so exactly with the change from a region of summer rains to arid desert that it is impossible to doubt that it is the peculiar topographical features of the country which account for the extraordinary climatic conditions, more particularly their startling changes.

West of T'sam-p'u-t'ong rises the great snow mountain of Ke-ni-ch'un-pu, separating the Salween from the headwaters of the 'Nmai-kha, and at the same time abruptly putting an end to the Salween jungles by depriving them of their vital rainfall.

To the east, between the Salween and the Mekong, rises the still more remarkable K'a-gur-pu, which is, as we have seen, continued northwards in a chain of high peaks to Ta-miu in Tibet.

Eastwards again we have a nameless snow mountain between the Mekong and the Yang-tze, continued, not northwards, but southwards by the snowy peaks of Pei-ma-shan, though it is only the scanty precipitation on this stupendous ridge which prevents many other peaks from being clad with eternal snow.

The really important rain-screen, then, is the K'a-gur-pu *massif,* since it is owing to the presence of this overwhelming mountain, which divides with Ke-ni-ch'un-pu the lion's share of the rains on this, the eastern limit of the monsoon, that the Mekong-Yang-tze watershed is entirely changed in character.

South of T'sam-p'u-t'ong there is no really big barrier west of the Salween ; ridge beyond ridge there is still, right away to the plains of Assam, but not one of them is sufficiently high to screen the country beyond from the effects of the monsoon and seriously affect the rainfall in the Salween valley.

But the Mekong rift is so extraordinarily narrow that the rains seem to pass right over it and precipitate themselves on the mountains and plains further east, so that this river continues to grind its way between dry bare cliffs. It is only for a stretch of about fifty miles south of Yangtsa, which may be called the rain-belt, that the Mekong gorges support anything approaching forest, and even the luxuriant vegetation occupying the gullies of the middle Mekong owes its existence, not to a more copious rainfall, but to the heavy night dews in a region of intense radiation, which are here protected from being immediately lapped up as soon as the sun gets into the valley. But as we are now wandering somewhat beyond the Land of Deep Corrosions, it will be unnecessary to pursue this subject further.

The rain-belt immediately succeeding the arid region is however something of a puzzle. Were it continued south-wards, the valley receiving a heavier and heavier rainfall as the dividing ridges to the west became lower, there would be no difficulty in understanding it. But it is not, for, as already stated, a second arid region begins outside the limits of the country immediately under discussion, and the term 'arid region' is henceforth confined to the intensely dry valleys lying between the snow-clad ridges soon destined to lose themselves in Tibet.

The effect of this scanty rainfall in the arid region, which can hardly exceed five inches annually, is greatly aggravated by the local winds, which throughout the summer blow up all three valleys with the regularity of the trade-winds, setting in soon after mid-day and blowing themselves out before midnight. This desiccating wind is due to the cold air pouring down from the high mountains immediately overhead to fill the partial vacuum caused by the intense heating of the narrow valley during the day; and the fact that it is almost invariably an up-valley wind may be ascribed to the fact that the gorges contract more and more as we proceed northwards, so that the vacuum becomes more and more complete in this direction.

I have travelled in the arid regions of both the Salween and Mekong valleys during the summer months beneath a winding ribbon of blue sky which followed the course of the river as a canopy might, and watched the clouds thickening on the mountains immediately to east and west till their summits were hidden in blinding rain-storms. Once at Yang-tsa on the Mekong, which almost exactly marks the junction between the rain-belt and the arid region in this valley, I watched a curious spectacle. At seven o'clock in the morning there was blue sky overhead extending northwards as far as the eye could reach, but the clouds were gathering in the south and away down the valley it was snowing heavily on the mountains, their summits being completely concealed. A furious wind was raging up the valley at the time. Very slowly the clouds sailed up to the assault, and an hour later there were broken masses of cloud just south of us. By ten o'clock there were puffs of cumulus almost overhead ; but struggle as they would they were unable to cross what seemed a physical barrier forcibly keeping them back. To the north the sky was blue and cloudless, and it was mid-day when a few scattered rain-drops fell at Yang-tsa. In the afternoon I walked some miles down the river, finding it heavily clouded over and a continuous drizzle falling, but the sky was still blue in the north though a few puffs of cloud had at last succeeded in crossing the dividing line, and were rapidly dwindling even as they triumphed. By mid-afternoon more clouds had forced the barrier and were concentrating themselves over the mountains on either side of the valley, but blue sky still maintained the position intact above the river.

This phenomenon served as an excellent illustration of the part played by these parallel ridges in determining the rainfall in the valleys to north and south of the rain-screen.

I have already alluded to the great altitude of the snow-line east of the Mekong, and also to the very considerable elevation of the peaks and passes on that watershed. There can be no doubt that, did the Mekong-Yang-tze ridge at present receive as heavy a rainfall as does the Mekong-Salween ridge, there would be snow-fields and glaciers on the former where none now exist. This is an important consideration, because after many climbs amongst these mountains I became convinced that there had actually been glaciers in many of the valleys. The valleys which descend immediately from the watershed are in the first place all hanging valleys, and at their upper extremities is always to be found in place of a gradual ascent growing steeper and steeper towards the cirque, a peculiar 'tread and riser' structure, the floor ascending in three or four big steps to the valley head, so that the streams alternately wander sluggishly in divided bands of water through a flat region of sand and shingle, and presently come tumbling in a cascade over heaps of boulders to the next level. Small lakes, sometimes occupying obvious rock-basins of considerable depth, occur in almost every valley, and in the bigger hanging valleys there may be as many as four or five, one at each level. In the smaller valleys there is commonly one only at the foot of the screes surrounding the valley head, where confused heaps of angular rock-fragments are piled indiscriminately at the base of the crags, and mounds of scree material, which might be lateral moraines, occur here and there bounding the valleys. Of striae or perched blocks however I could find no definite trace.

One further piece of indirect evidence may be adduced in favour of previous glaciation. I have spoken of a big snow mountain on the Mekong-Yang-tze watershed to the south-east of A-tun-tsi, known to the Chinese as Pei-ma-shan, and from the Tung-chiu-ling road I obtained good views of its glaciers. Their bottle-shaped snouts and the fact that the terminal moraine was, in one case at least, some distance from the foot of the glacier, indicated that they were in a state of retreat, and moreover that they were retreating rapidly; no very big snow-fields were visible, and a great deal of bare rock was exposed. Evidently the glaciers of Pei-ma-shan are mere shrunken remnants of their former selves.

There are on this divide numerous peaks rising to 18,000 feet and more, but all save the two referred to are clear of snow for perhaps three months in the year. Were the rainfall doubled this emphatically would not be the case. If then glaciers, which have since disappeared, did once fill these valleys, their disappearance is almost certainly to be traced to a reduction of the rainfall.

Now if we imagine the high Mekong-Salween watershed swept away, the next ridge to the east, namely the Mekong-Yang-tze watershed, would receive the heavy rains which under existing conditions deluge the former. As previously stated, snow lay deep on the Mekong-Salween divide at

13,000 feet in June and at 12,000 feet in November, so that there is good reason to think that the snow-line on this side does not exceed 16,000 feet.

For these reasons I regard the Mekong-Salween watershed as having been raised—at all events to its present very considerable elevation—subsequent to the Mekong-Yang-tze watershed, and this at once suggests that the lateral pressure which I assume to have squeezed the entire region into its remarkable form came from the west, not from the east, since the last ridge to be raised up, granting a certain degree of rigidity to have been attained further east, would naturally be on the side from which the pressure acted.

It is not unlikely that a comprehensive survey of the distribution of plants in Western China, particularly of the truly alpine species, together with a careful comparison of the floras on the dividing ridges, would lead to useful results as regards the history of these rivers. It is an interesting point of view from which to look at the problem, but we need a much more extensive knowledge of the facts than we have at present, before drawing definite conclusions, though certain broad principles are so strikingly illustrated that they are worth referring to.

The outstanding feature of the alpine flora is that it is essentially—indeed entirely—a North Temperate one, agreeing with Drude's Northern Floral Region.

Characteristic orders which are well represented are Ranunculaceae, Papaveraceae, Rosaceae, Saxifragaceae, Umbelliferae, Compositae, Primulaceae, Gentianaceae, Scrophulariaceae, and Liliaceae, while the following genera present a richness of species I have never seen equalled elsewhere; Rhododendron, Gentiana, Saxifraga, Meconopsis, Primula, Pedicularis, and Corydalis. The formations characteristic of the Northern Glacial Zone and the Northern Zone of cold winters (i.e. the Arctic, and large parts of Europe) are also typical of these mountains, tundra being represented by the high alpine flora with 'cushion' plants, forest by conifers and catkin-bearing trees, grass-land by alpine meadow and turf, heath by dwarf rhododendron. The vegetative season lasts about four months, from May to August on the Mekong-Salween divide, and from June to September on the Mekong-Yang-tze divide.

Plate XXXIX

The Salween forests in summer,
Mekong-Salween Divide, 8000 feet

The Salween forests in winter,
Mekong-Salween Divide, 12,000 feet

In this connection it must be remembered that our high mountain chains are in direct communication with the Tibetan plateau and form ideal lines of migration.

If now we compare the flora of the Mekong-Salween divide with that of the Mekong-Yang-tze divide, we arrive at an apparently anomalous conclusion, for a traveller crossing these two watersheds in the summer would almost unhesitatingly pronounce the alpine flora of the Mekong-Salween divide to be the richer of the two. Yet I do not believe this first impression would be correct, for closer examination seems to me to reveal the fact that, while the Mekong-Salween divide is richer in genera, the Mekong-Yang-tze divide is richer in species, but that there is little difference in the sum total of species on the two ridges. Even this, however, might seem to argue against the Mekong-Salween ridge having come into prominence subsequent to the Mekong-Yang-tze ridge, since it requires a longer period of time for a varying group of plants to acquire generic rank than it does for them to be classed as distinct species ; but it is quite a gratuitous assumption to suppose that they have attained generic rank *in situ*, whereas there is every reason from the nature of the case to think that the high alpine flora of the Mekong-Yang-tze divide has greatly enriched itself *in situ*.

Again, if the Mekong-Yang-tze watershed did actually, ages ago, receive a greater rainfall which has since been reduced, it is natural to suppose that many plants adapted to flourish in an extremely wet climate would by now have disappeared or become greatly reduced in numbers ; and the few plants to which I paid attention, common to the two mountain chains, rather bear out this supposition by their distribution.

The sulphur-yellow *Meconopsis integrifolia* for example, a plant which flourishes in a wet climate, occurs very sparsely on the Mekong-Yang-tze ridge, being confined to a few favourable localities, whereas whole meadows of it are to be seen on the Mekong-Salween ridge. The plant is very common to the north and east, on the Ssü-chuan mountains, whence it is reasonable to infer that it has come down these ridges, having since almost disappeared from the Mekong-Yang-tze divide, though it is of course possible that we find there, not a remnant of a former extensively distributed species, but a plant which is trying to establish itself.   Such cases suggest lines of investigation, but require a much wider acquaintance with facts to justify conclusions of any importance

It is significant that while those plants common to both divides are unusually rare on the Yang-tze side and extremely abundant on the Salween side of the Mekong, this difference tends to disappear the nearer we approach the limit of plants (which, by the way, is very high on the Mekong-Yang-tze watershed) for the reason that during the rainy season there are continuous mists hanging over the peaks, sufficient to keep the atmosphere saturated even when no rain is actually falling, while the heavy dews experienced at great elevations may also have something to do with it.

Thus several species of *Primula* (e.g. *P. bella*) occur almost as commonly at 15,000—16,000 feet on the Mekong-Yangtze divide as they do at 13,000—14,000 feet on the Mekong-Salween divide, which again suggests not only that the flora was originally the same on both sides (and certainly it would be a most remarkable circumstance if it were not!), but also that it is the rainfall which has largely determined existing differences in the flora of the two ridges ; such differences being far more noticeable in the forest belt, the lower limit of which is defined entirely by the rainfall, than in the alpine belt.

Similarly the very conspicuous fact that the flora of the Mekong-Salween divide is essentially a summer flora, attaining its maximum development during the months of June and July, while that of the Mekong-Yangtze divide is equally an autumn flora, attaining its maximum development between August and October, may be attributed to the seasonal distribution of rain in the two cases, as well as to the fact that snow is beginning to fall on the Mekong-Salween divide early in October.

The comparative preponderance of species over genera on the Mekong-Yang-tze divide, especially in the two genera *Saxifraga* and *Gentiana*, which occur in enormous numbers, may be adduced as evidence for a *changing* climate.   A very large extent of territory is available to plants on the one watershed which is denied them on the other, and one would expect that, as the glaciers retreated, the flora would gradually take advantage of the opportunity to occupy new territory, which would at once bring it into contact with new physical conditions.

The whole idea of attacking the geological problem from a botanical point of view, however, opens up such a wide field for investigation that it is useless to pursue it as an aimless speculation without marshalling an enormous array of facts in support of this or that contention. I have merely made certain suggestions based upon limited observations, the accuracy of which is in some cases unfortunately open to criticism, and I must leave it to others to say whether they are justified or the reverse.

Here, then, I will conclude this brief survey of some scientific problems which visions of the Land of Deep Corrosions at once conjure up. Convinced as I am that with its wonderful wealth of alpine flowers, its numerous wild animals, its strange tribes, and its complex structure it is one of the most fascinating regions of Asia, I believe I should be content to wander over it for years. To climb its rugged peaks, and tramp its deep snows, to fight its storms of wind and rain, to roam in the warmth of its deep gorges within sight and sound of its roaring rivers, and above all to mingle with its hardy tribesmen, is to feel the blood coursing through the veins, every nerve steady, every muscle taut.

APPENDIX I

The following is a preliminary and incomplete list of plants collected during the expedition. Of those marked † seeds were obtained; they are now being grown by Bees Ltd. in their Cheshire nurseries, and many of them will shortly be on the market. Those marked with an asterisk are new species.

I am indebted for this list to Professor Bayley Balfour, F.R.S., and Mr W. W. Smith, who are also growing many of the plants from seed at the Edinburgh Botanic Gardens, where the dried material may also be seen.

1. †*Aconitum Hookeri*, Stapf. f. gibbo nectarii magis producto.
2. *Anemone obtusiloba*, Don. var.
3. † „ *rupicola*, Camb.
4. *Callianthemum cachemirianum*, Camb.
5. *Caltha palustris*, Linn.
6. †*Isopyrum grandiflorum*, Fisch.
7. *Oxygraphis glacialis*, Bunge.
8. †*Podophyllum Emodi*, Wall.
9. *Ranunculus hirtellus*, Royle.
10. *Souliea vaginata* (Maxim.), Franch.
11. †*Trollius patulus*, Salisb. var.
12. „ *pumilus*, Don.
13. *Arabis alpina*, Linn., var. *rubrocalyx*, Franch.
14. *Braya rubicunda*, Franch.
15. *Cardamine granulifera* (Franch.), Diels.
16. *Braya sinensis*, Hemsl.
17. ***Cardamine verticillata*, Jeff. et W. W. Sm. Sp. nov.
18. *Draba alpina*, Linn. var.
19. *Thlaspi yunnanense*, Franch.
20. *Arenaria Forrestii*, Diels.
21. „ *polytrichoides*, Edgew.
22. „ *Delavayi*, Franch.
23. *Lychnis nigrescens*, Edgew.
24. *†*Silene rosaeflora*, Kingdon Ward. Sp. nov.
25. *Geranium Pylzowianum*, Maxim.
26. *Astragalus yunnanensis*, Franch. var.
27. *Gueldenstaedtia yunnanensis*, Franch.

28.   *Hedysarum sikkimense*, Benth.
29.   *Caragana crassicaulis*, Benth.
30.   *Astragalus wolgensis*, Bunge.
31.   *Thermopsis inflata*, Cambess.
32.   †*Potentilla articulata*, Franch.
33.   †   „   *fruticosa*, Linn., var. *armerioides*, Hook. f.
34.   †   „   *peduncularis*, Don. forma.
35.   †   „   *Saundersiana*, Royale, var. *Jacquemontii*, Fr.
36.   *Sedum* sp.
37.   *Sedum* sp.
38.   †*Saxifraga atrata*, Engl.
39.   „   *cardiophylla*, Franch.
40.   „   *micrantha*, Edgew. forma.
41.   †   „   *nigroglandulosa*, Engl. et Irmscher.
42.   „   *sibirica*, Linn.
43.   *†   „   *atuntsiensis*, W. W. Sm. Sp. nov.
44.   *†   „   *consanguinea*, W. W. Sm. Sp. nov.
45.   *†   „   *flexilis*, W. W. Sm. Sp. nov.
46.   *†   „   *Wardii*, W. W. Sm. Sp. nov.
47.   *†   „   *finitima*, W. W. Sm. Sp. nov.
48.   †   „   *chrysanthoides*, Engl. et Irmscher.
49.   †   „   *purpurascens* (H. f. et T.), var. *Delavayi* (Fr.), Engl. et Irmscher.
50.   *Epilobium angustifolium*, Linn.
51.   *Pleurospermum foetens*, Franch.
52.   *Trachydium chloroleucum*, Diels.
53.   *†*Meconopsis Wardii*, Prain.
54.   *Parrya* sp. (Incomplete.)
55.   *Corydalis Balfouriana*, Diels.
56.   „   sp.
57.   „   sp.
58.   „   sp.
59.   „   *pulchella*, Franch.
60.   „   sp.
61.   „   sp.
62.   „   sp.
63.   „   sp.
64.   *Morina Bulleyana*, G. Forrest et Diels.
65.   *Anaphalis xylorhiza*, Schultz-Bip.
66.   *Tanacetum tenuifolium*, J. Gay.
67.   †*Cremanthodium Decaisnei*, C. B. Clarke.
68.   †   „   *rhodocephalum*, Diels.
69.   *Crepis rosularis*, Diels.
70.   †*Lactuca Souliei*, Franch.

71. *Crepis Umbrella*, Franch.
72. **Saussurea quercifolia*, W. W. Sm. Sp. nov.
73. * „ *loriformis*, W. W. Sm. Sp. nov.
74. **†Cremanthodium bupleurifolium*, W. W. Sm. Sp. nov.
75. †*Codonopsis convolvulacea*, Kurz.
76. „ *tubulosa*, Komarow.
77. †*Pyrola atropurpurea*, Franch.
78. *Diapensia himalaica*, H. f. et T.
79. *Androsace chamaejasme*, Host.
80. „ *Delavayi*, Franch.
81. † „ *spinulifera*, Franch.
82. †*Primula bella*, Franch.
83. † „ *brevifolia*, G. Forrest.
84. † „ *calliantha*, Franch.
85. † „ *dryadifolia*, Franch.
86. † „ *lichiangensis*, G. Forrest.
87. † „ *septemloba*, Franch.
88. † „ *pulchella*, Franch.
89. † „ *nivalis*, Pall. forma.
90. † „ *sikkimensis*, Hook. forma.
91. † „ *Watsoni*, Dunn.
92. „ *sphaerocephala*, Balf. f.
93. „ *cernua*, Franch.
94. „ *Giraldiana*, Pax.
95. „ *silaensis*, Petitm.
96. * „ *vernicosa*, Kingdon Ward. Sp. nov.
97. „ *serratifolia*, Franch.
98. † „ *sibirica*, Jacq.
99. † „ *sonchifolia*, Franch.
100. † „ *Franchetii*, Pax.
101. † „ *vittata*, Franch.
102. *Gentiana aprica*, Dcne.
103. † „ *ornata*, Wall.
104. † „ *heptaphylla*, Balf. f. et G. Forrest.
105. *† „ *Wardii*, W. W. Sm. Sp. nov.
106. „ *Georgei*, Diels.
107. „ *cyananthiflora*, Franch.
108. *Pleurogyne oreocharis*, Diels.
109. *Gentiana decorata*, Diels.
110. *† „ *atuntsiensis*. Sp. nov.
111. †*Cynoglossum amabile*, Stapf. et Drummond.
112. *Tylophora* sp.
113. *Eritrichium* sp.
114. *Myosotis Hookeri*, Clarke.

115.	*Mandragora caulescens*, Clarke.	
116.	*Lancea tibetica*, Hook. f. et Thoms.	
117.	*Pedicularis atuntsiensis*, Bonati. Sp. nov.	
118.	„ *cephalantha*, Franch.	
119.	„ *cibaria*, Maxim.	
120.	„ *cranolopha*, Maxim.	
121.	„ *densispica*, Franch.	
122.	„ *Elwesii*, Hook. f.	
123.	„ *gyrorhyncha*, Franch.	
124.	„ *labellata*, Jacquem.	
125.	„ *lachnoglossa*, Hook. f.	
126.	„ *leiandra*, Franch.?	
127.	„ *lineata*, Franch.	
128.	„ *longiflora*, Rud.	
129.	„ *macrosiphon*, Franch.	
130.	„ *Oederi*, Vahl.	
131.	„ *Przewalskii*, Maxim.	
132.	* „ *pseudo-ingens*, Bonati. Sp. nov.	
133.	„ *rex*, Clarke.	
134.	„ *rupicola*, Franch.	
135.	„ *siphonantha*, Don.	
136.	„ *strobilacea*, Franch.	
137.	„ *superba*, Franch.	
138.	„ *yunnanensis*, Franch.	
139.	†*Didissandra lanuginosa*, Clarke, var. *lancifolia*, Franch.	
140.	*Nepeta complanata*, Dunn. Sp. nov.	
141.	†*Phlomis rotata*, Benth.	
142.	*Scrophulariaceae* (undetermined).	
143.	*Polygonum Forrestii*, Diels.	
144.	„ *sphaerostachyum*, Meissn. forma.	
145.	*Euphorbia Stracheyi*, Boiss.	
146.	*Cephalanthera falcata*, Lindl.	
147.	*Cypripedium Wardii*, Rolfe. Sp. nov.	
148.	„ *tibeticum*, King.	
149.	„ *arietinum*, Raf.	
150.	„ *guttatum*, Schwarz.	
151.	*Listera Wardii*, Rolfe. Sp. nov.	
152.	*Oreorchis foliosa*, Lindl.	
153.	*Nervilia tibetensis*, Rolfe. Sp. nov.	
154.	*Roscoea alpina*, Royle.	
155.	*Allium monadelphum*, Turcz. var.	
156.	*Fritillaria Souliei*, Franch.	
157.	„ *Delavayi*, Franch. var.	
158.	†*Lilium lophophorum*, Franch. var.	

159.   *Lilium lophophorum*, Franch. var.
160.   *Lloydia Forrestii*, Diels.
161.   „   *serotina*, Rchb., var. *unifolia*, Franch.
162.   „   *tibetica*, Baker.
163.   *Ophiopogon Wallichianum*, Hook. f.
164.   *Streptopus amplexifolius*, DC.
165.   *Liliaceae* (undetermined).
166.   *Iris kumaonensis*, Wallich.
167.   *Juncus longistamineus*, Camus?
168.   „   *sikkimensis*, Hook. f.
169.   *Botrychium Lunaria*, Linn. forma multifida.
170.   *Woodsia hyperborea*, R. Br.
171.   *Morina Delavayi*, Franch.
172.   *†*Androsace Wardii*, W. W. Sm. Sp. nov.
173.   †*Meconopsis speciosa*, Prain.
174.   †   „   *rudis*, Prain.
175.   †   „   *pseudointegrifolia*, Prain.
176.   †   „   *integrifolia*, Franch.
177.   *Hedysarum* sp.
178.   †*Primula pulchelloides*, G. Forrest.
179.   *Campanula colorata*.
180.   †*Amphicome arguta*.
181.   †*Incarvillea* sp.
182.   †*Oxalis* sp.
183.   †*Eremurus chinensis*.
184.   †*Stellera* sp.
185.   †*Paeonia Delavayi*.
186.   †*Senecio dictyoneurus*.
187.   †*Aster Delavayi*.
188.   †*Rodgersia* sp.
189.   †*Wikstroema* sp.
190.   †*Sophora viciifolia*.
191.   †*Rosa* sp.
192.   †*Rubus* sp.
193.   †*Rhododendron* sp.
194.   †*Androsace Bulleyana*.
195.   †*Lilium giganteum*.
196.   †*Aconitum* sp.
197.   †*Delphinium yunnanense*.
198.   †*Clematis montana*.
199.   †   „   *Delavayi*.
200.   †   „   *splendens*.

APPENDIX II

The following is a list of the small mammals collected during the expedition. The specimens, which are now at the Natural History Museum, were identified and described by Mr Oldfield Thomas, F.R.S. (see *Annals and Magazine of Natural History*, Ser. 8, Vol. IX. May 1912) to whom I am indebted for permission to append the list.

Those with an asterisk are new species.

1. *Scaptonyx fusicaudatus affinis.* Subsp. nov.

   ♂. Trapped on mossy bank in *Abies* forest, 14,000 feet.

2. *Marmota robusta.*

   ♀. Young. Shot at 15,000 feet.

3. *Epimys confucianus.*

   ♀. 11,500 feet.

4. *Apodemus speciosus latronum.*

   ♂ and ♀. 12,000 feet. The commonest field mouse round A-tun-tsi.

5. *Apodemus chevrieri.*

   ♀. 12,000 feet.

6. *Microtus irene.*

   ♂ and ♀. 15,000 feet.

7. **Microtus wardi.* Sp. nov.

   ♂. 13,000 feet, Mekong-Salween divide. This vole is trapped and eaten by the Lutzu of the Salween valley.

8. **Microtus custos.* Sp. nov.

   ♂ and ♀. 12,000 feet. The commonest vole round A-tun-tsi.

9. *Ochotoma roylei chinensis.*

   ♀. 16,000 feet. The highest mammal seen, and only discovered a few months previously by Captain Bailey at Ta-tsien-lu.

INDEX

CAMBRIDGE: PRINTED BY JOHN CLAY, M.A. AT THE UNIVERSITY PRESS

Printed in the United States
By Bookmasters